Identification of Distributed Systems

Modern Analytic *and* Computational Methods *in* Science *and* Mathematics

A GROUP OF MONOGRAPHS AND ADVANCED TEXTBOOKS

Richard Bellman, EDITOR
University of Southern California

Published

1. R. E. Bellman, Harriet H. Kagiwada, R. E. Kalaba, and Marcia C. Prestrud, Invariant Imbedding and Radiative Transfer in Slabs of Finite Thickness, 1963

2. R. E. Bellman, Harriet H. Kagiwada, and Marcia C. Prestrud, Invariant Imbedding and Time-Dependent Transport Processes, 1964

3. R. E. Bellman and R. E. Kalaba, Quasilinearization and Nonlinear Boundary-Value Problems, 1965

4. R. E. Bellman, R. E. Kalaba, and Jo Ann Lockett, Numerical Inversion of the Laplace Transform: Applications to Biology, Economics, Engineering, and Physics, 1966

5. S. G. Mikhlin and K. L. Smolitskiy, Approximate Methods for Solution of Differential and Integral Equations, 1967

6. R. N. Adams and E. D. Denman, Wave Propagation and Turbulent Media, 1966

7. R. L. Stratonovich, Conditional Markov Processes and Their Application to the Theory of Optimal Control, 1968

8. A. G. Ivakhnenko and V. G. Lapa, Cybernetics and Forecasting Techniques, 1967

9. G. A. Chebotarev, Analytical and Numerical Methods of Celestial Mechanics, 1967

10. S. F. Feshchenko, N. I. Shkil', and L. D. Nikolenko, Asymptopic Methods in the Theory of Linear Differential Equations, 1967

11. A. G. Butkovskiy, Distributed Control Systems, 1969

12. R. E. Larson, State Increment Dynamic Programming, 1968

13. J. Kowalik and M. R. Osborne, Methods for Unconstrained Optimization Problems, 1968

14. S. J. Yakowitz, Mathematics of Adaptive Control Processes, 1969

15. S. K. Srinivasan, Stochastic Theory and Cascade Processes, 1969

16. D. U. von Rosenberg, Methods for the Numerical Solution of Partial Differential Equations, 1969

17. R. B. Banerji, Theory of Problem Solving: An Approach to Artificial Intelligence, 1969

18. R. Lattès and J.-L. Lions, The Method of Quasi-Reversibility: Applications to Partial Differential Equations. Translated from the French edition and edited by Richard Bellman, 1969

19. D. G. B. Edelen, Nonlocal Variations and Local Invariance of Fields, 1969

20. J. R. Radbill and G. A. McCue, Quasilinearization and Nonlinear Problems in Fluid and Orbital Mechanics, 1970

21. W. Squire, Integration for Engineers and Scientists, 1970

23. T. Hacker, Flight Stability and Control, 1970

24. D. H. Jacobson and D. Q. Mayne, Differential Dynamic Programming, 1970

25. H. Mine and S. Osaki, Markovian Decision Processes, 1970

26. W. Sierpiński, 250 Problems in Elementary Number Theory, 1970

27. E. D. Denman, Coupled Modes in Plasmas, Elastic Media, and Parametric Amplifiers, 1970

29. G. A. Phillipson, Identification of Distributed Systems, 1971

30. D. H. Moore, Heaviside Operational Calculus: An Elementary Foundation, 1971

32. V. F. Demyanov and A. M. Rubinov, Approximate Methods in Optimization Problems, 1970

34. C. J. Mode, Multitype Branching Processes: Theory and Applications, 1971

In Preparation

22. T. Parthasarathy and T. E. S. Raghavan, Some Topics in Two-Person Games

28. F. A. Northover, Applied Diffraction Theory

31. S. M. Roberts and J. S. Shipman, Two-Point Boundary Value Problems: Shooting Methods

33. S. K. Srinivasan and R. Vasudevan, Introduction to Random Differential Equations and Their Applications

35. R. Tomović and M. Vukobratović, General Sensitivity Theory

36. J. G. Krzyż, Problem Book in Artificial Functions

37. W. T. Tutte, Introduction to the Theory of Matroids

38. B. W. Rust and W. R. Burrus, Mathematical Programming and the Numerical Solution of Linear Equations

Identification of Distributed Systems

G. A. Phillipson
Shell Development Company
Houston, Texas

American Elsevier
Publishing Company, Inc.
NEW YORK · 1971

AMERICAN ELSEVIER PUBLISHING COMPANY, INC.
52 Vanderbilt Avenue, New York, N.Y. 10017

ELSEVIER PUBLISHING COMPANY, LTD.
Barking, Essex, England

ELSEVIER PUBLISHING COMPANY
335 Jan Van Galenstraat, P.O. Box 211
Amsterdam, The Netherlands

International Standard Book Number 0-444-00091-7

Library of Congress Card Number 71-110381

Printed in the United States of America

Contents

5. Solution of the State Identification Problem

6. Summary and Extensions

PREFACE

This monograph is an outgrowth of a doctoral thesis written at Case Institute of Technology in the summer of 1968. As implied by the title, it treats the state identification problem associated with two elementary classes of linear distributed systems, namely, parabolic (diffusion) and second order hyperbolic (wave) systems. The treatment is rigorous. Such treatment of the elementary systems serves two major purposes. First, the restriction of attention to these elementary systems makes strong characterization of the solution possible; that is, few assumptions of a restrictive nature need to be made. It is unnecessary to point out that the customary treatment of a canonical set of equations in the lumped theory has no parallel in the distributed context, except in a very formal and almost uninteresting sense. A more objectionable, though superficially appealing route, is to lump the distributed systems and then apply the lumped theory. Of course, in some instances, such approaches can be made respectable. A second purpose is the provision of a foundation for the study of more complex distributed systems, perhaps by linearization (if appropriate) or other approximations (such as the method of "quasi-reversibility" of Lions). This foundation is less extensive than, but shares strong similarities with, the "linear-quadratic" problem so familiar in the lumped setting.

The approach to the solution of the state identification problem adopted here is a deterministic, variational one. The solutions obtained in this way can be given (and are) a "stochastic respectability"—with 20/20 hindsight. This procedure avoids the introduction of considerable mathematical structure while maintaining the rigorous approach. Researchers in optimal control of distributed systems will find interest in and application of some of the results, particularly in the approach to the solution of the necessary conditions for optimality.

The numerical solution of problems associated with distributed systems is an important and most often neglected area of research in the current control theory literature. Two numerical techniques which are familiar to researchers in the lumped control theory for the solution of the state identification problem are presented. Specifically, the treatment is exclusively in terms of partial differential equations (or partial integro-differential equations). Numerical solutions of these equations are obtained with tools no more sophisticated than those applicable to a set of ordinary differential equations. In particular, a modified Galerkin approximation procedure is adopted for the solution of these equations. In one instance, however (in the context of solving a Riccati-like equation familiar in the lumped theory), solutions applicable to solving Fredholm equations of the second kind are suggested. There is an attempt to cast (theatrically speaking) the "theoretical" and "numerical" problems as stars with equal billing. The proposed numerical approach is demonstrated by an example chosen to illustrate the features of the state identification problem.

G. A. PHILLIPSON

Houston, Texas
Fall 1970

ACKNOWLEDGEMENTS

I am indebted to J. D. Pearson for the initial ideas, S. K. Mitter for tireless guidance, J. L. Lions for midcourse correction, and the environment at Case Institute of Technology for constructive interaction and support.

CHAPTER 1

INTRODUCTION

A.0 THE IDENTIFICATION PROBLEM IN CONTEXT

The intelligent control of the behavior of some abstract system experiencing constraining influences from its environment is an all-pervading theme in the engineering experience. Frequently the exertion of this "intelligent control" must be performed in the face of inexact and incomplete data pertaining to the evolution of the system in its environment. This is the context in which the identification problem arises. Stated simply, the problem or goal of identification is to provide the decision maker (intelligent controller) with a summary of the position (in a generalized coordinate space) of the system, which is optimal (in some sense), with respect to the available data pertaining to environmental interaction and system evolution.

We are concerned with systems whose evolution processes are described by a class of partial differential equations, so that, in the context of the preceding discussion, the "position" of the system is "summarized" by a specification of the solution to the given partial differential equation. This specification is contingent upon a knowledge of the environmental interaction processes. That is, the generation of the solution to the partial differential equation requires a statement of the:

(i) initial condition(s);

(ii) environmental forcing terms, which include the boundary conditions.

It is assumed that the initial conditions (i) with which the system begins its evolution, as well as the environmental interaction processes (ii), are not known with precision. However, the "available data" include inexact measurements of (i) and (ii) and, in addition, provides inexact and possibly incomplete measurements of the state of evolution of the system. Thus the identification problems considered in this study are the following:

P: Determine, on the basis of the available data, *an estimate* of the true initial and boundary conditions associated with a given partial differential equation which is, in some sense, optimal with respect to the given data. Problem P is the so-called state identification problem, as distinguished from the system identification or parameter identification problems, wherein the objectives are to specify the system evolution process or to obtain estimates of parameters influencing the evolution of some known system, respectively.

The basis for selecting the estimates of the boundary and initial conditions associated with a given partial differential equation, that is, the criterion of optimality, is that of "least squares." To be more precise, we mean the following:

Given:

(i) the measurement data, which we denote here by Z, and

(ii) an (arbitrary) solution of the partial differential equation, denoted here by $Y(\mathbf{v})$, where $\mathbf{v}$ is an arbitrary *estimate* of the true initial state and boundary conditions, then

obtain:

(i) $\mathbf{v}$ which extremizes the *error functional*

$$J(\mathbf{v}) = \|Z - Y(\mathbf{v})\|^2,$$

where $\|\cdot\|^2$ is some appropriate squared metric.

Evidently then, the identification problem as treated in this study, when stripped of all the descriptive embellishment, is a variational problem, that of characterizing extremals to a given functional, constrained by a partial differential equation. The variational framework adopted in this context is uncommon, though Lions [1] and Balakrishnan and Lions [2] have considered some state identification problems in precisely this way. Owing to the variational framework of the problem, the results of other investigators in the area of optimal control of distributed systems is relevant. In that context we should mention the pioneering work of Wang [3] in the U.S.A. and Butkovskii [4,5] in the U.S.S.R. Both investigators presented a "maximum principle" for a very general class of partial differential equations and "cost" functionals. These principles were (and are) weakened by certain formal assumptions which were necessitated (in part) by the generality of the class of variational problems.

A salient difference in the statement of the maximum principles of Wang and Butkovskii is that the latter used an integral equation approach to the "solution," whereas the former applied the dynamic programming formalism. Because of the algorithmic character of this formalism, it is used almost exclusively in contemporary studies of the optimal control problem for distributed systems. A feature of solutions obtained in this way is that they are formal, that is, the principle may characterize a nonexistent solution.

It has been apparent to some investigators (principally Lions [6], Erzberger and Kim [7]) that a reduction in generality of the class of distributed systems treated would lead to a strengthened characterization of extremals of the associated variational problem. Erzberger and Kim treat a diffusion system via dynamic programming, while Lions generates a new maximum principle, giving necessary and sufficient conditions to the solution of certain specialized variational problems associated with general "diffusion" and "wave" systems.

We use this principle and obtain new results for the identification problem associated with diffusion and wave-type systems, phrased as a variational problem.

Another approach to the distributed variational problem has been taken. There the distributed systems are first "lumped" and then, to this equivalent lumped system the established maximum principle of L. S. Pontriagin is applied. This approach, while intuitively appealing, has mathematical pitfalls which render the results obtained in this way unsound. See, for example, [8].

We should say that numerical techniques for the solution of these variational problems are conspicuous by their absence, although [7] does present a numerical solution. Two numerical methods for the solution of the variational problem are presented here, one of which is novel. In addition, a new numerical technique for the approximation of solutions to distributed systems using spline interpolation is also considered. Before presenting a review of the monograph content, we remark first on the identification criterion adopted (namely, least squares) and the connection with another approach.

A.1 STOCHASTIC RESPECTABILITY

The identification problem, as introduced, is customarily given a stochastic treatment. In that context the error associated with the measurement data Z is endowed with statistical properties. That is, the error is considered to be a statistically random variable whose values are "distributed" in a known way. The state identification or filtering problem, as it is called in this context, is to determine the *a posteriori* probability density of the state, given the measurements Z. Equivalently, the definition of the sufficient statistics of this distribution solves the filtering problem. In a sense these sufficient statistics are optimal estimates of the system state, since knowledge of them enables a statement of the "maximum likelihood estimate" or any other statistical estimate.

Under special statistical hypothesis on the error processes, namely, that they be purely random with Gaussian probability density and in addition, are additive; that is,

$$Z = Y(\mathbf{u}) + E,$$

where $\mathbf{u}$ is the true "state of nature" and E is the error process, then, if the system state evolution process is also linear, the *a posteriori* density of the states is also Gaussian. We show in Chapter 3 that the filtered estimate (given in terms of the sufficient statistics of the Gaussian distribution of the states, the mean and variance) coincides with the "least squares" estimate. Thus under this special hypothesis the variational and stochastic approaches yield identical results, and the variational approach has "stochastic respectability."

The filtering problem introduced, except in the special case considered, is difficult. Indeed, even in the lumped case, where results have been obtained [9–11], solutions to the resulting equations characterizing the *a posteriori* probability density of the states must be approximated. In the distributed case, Falb [12] has recently obtained equations for the sufficient statistics of the pertinent probability density under the specialized hypotheses mentioned. A thorough and rigorous treatment of some aspects of the distributed identification problem from a stochastic point of view has been made by Bensoussan [13].

It was the knowledge of these difficulties and also the "stochastic respectability" of the least squares estimate which helped persuade us to take the variational approach.

A.2 DISTRIBUTED SYSTEMS

In Section A.0. we indicated that the systems of immediate concern are those whose states are described by partial differential equations. Such systems are commonly called distributed parameter systems or simply distributed. To be specific, the state evolution processes (that is, the differential equations) which are considered in this study are of two types:

(i) linear Parabolic partial differential equations and
(ii) linear second-order hyperbolic partial differential equations..

We give in Chapter 3 a precise definition of these equations, with an example (in each case) of the sort of physical phenomena characterized by them. Suffice it to say here that one can construct examples (admittedly idealized) from across the industrial spectrum—from glass making to aerospace. For example, the temperature of molten glass flowing slowly in a forehearth is described by an equation of type (i), whereas the displacements occasioned by dynamic loading on a slender airframe can be described by equations of type (ii). In the former example, temperature measurements are available at selected points along the spatial domain (obtained by pyrometer or some other device), whereas, in the latter case, strain gauge measurements at selected points on the airframe are reduced to yield the deflection data. In both cases the measurements are incomplete in the sense that the entire spatial profile is not available. Moreover, the measurements are inexact by virtue of inherent errors of measurement associated with transducing elements and also because of the measurement environment.

A.3 SUMMARY OF CONTENT

In Chapter 2, we consider the notation and theoretical preliminaries necessary to the mathematical statement of the identification problem, which, together with a precise statement of the class of systems considered, is given in Chapter 3. Chapter 3 also contains a demonstration of the stochastic respectability of the least squares estimate. In Chapter 4 a number of identification problems are posed, and their solutions are given in terms of a canonical set of partial differential equations. In Chapter 5 we consider two methods for solving this canonical set of equations. These solution techniques are applied to an illustrative numerical example, and the results are evaluated. Concluding remarks and suggestions for the solution of a parameter identification problem (using the framework provided in the preceding chapters) are given in Chapter 6.

CHAPTER 2

NOTATION, DEFINITIONS, THEORETICAL PRELIMINARIES

A.0 INTRODUCTION

This chapter contains the theoretical foundations on which the solution to the identification problem is constructed. We begin, in Section A.1, by consideration of the notation to be used in the sequel and the presentation of some pertinent definitions. In Section A.2 we give the properties of the solutions to two classes of partial differential equations. In Section A.3 a method for approximating the solution of partial differential equations, the Galerkin method, is presented along with some error estimates associated with the method. Section A.4 contains the fundamental theorems concerning the properties of extremals to quadratic functionals in a Hilbert space, which relate directly to the solution of the identification problem, appropriately phrased (Chapter 3, Section A.2). Those theorems are due to Lions and Stampaccia [14].

A.1 NOTATION AND DEFINITIONS

We shall consider functions defined on the following sets:

Ω, a simply connected, open set in R^r. Points of Ω are denoted by $x = (x_1 x_2, \ldots, x_r)$. Let t denote time, $t \in (0, T]$.

Γ, the boundary of Ω.

$\Sigma = \Gamma \times (0, T]$.

$Q = \Omega \times (0, T]$.

The set Q may be visualized, with the aid of Figure 2.1, as a cylindrical volume in $(r+1)$-dimensional Euclidean space enclosed within the sheath with "walls" Σ and "bottom" Ω.

We shall be concerned with equivalence classes of functions of a particular genre, namely, the (separable) Hilbert spaces. In particular, let $L^2(\Omega) = H$ denote the space (equivalence class) of real functions whose second powers are integrable for the measure $dx = (dx_1 dx_2 \ldots dx_r)$. Define the inner product and norm of elements $f, g \in H$:

$$\text{Inner product:} \quad (f, g)_{L^2(\Omega)} = \int_{\Omega} f(x)g(x)\,dx;$$

$$\text{Norm:} \quad \|f\|_{L^2(\Omega)} = \left[\int_{\Omega} f(x)^2\,dx\right]^{1/2}.$$

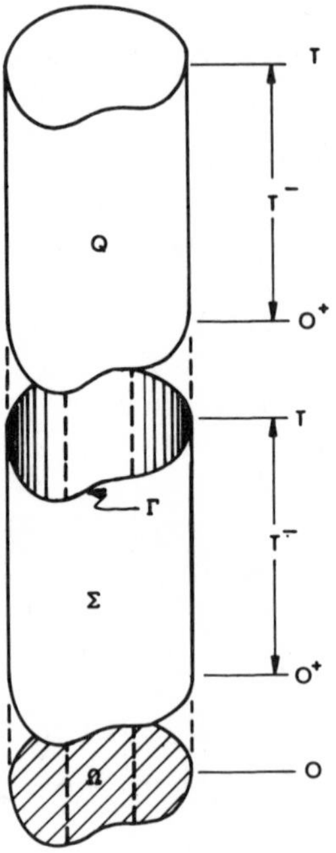

Figure 2.1 "Exploded" schematic of the domain considered.

Let $L^2(\Sigma)$ denote the space of real functions whose second powers are integrable for the measure $d\Sigma = (ds\,dt)$, where s is a point of Γ. Define the inner product and norm of elements $f, g \in L^2(\Sigma)$:

$$\text{Inner product:} \quad (f, g)_{L^2(\Sigma)} = \int_{\Sigma} f(s,t)g(s,t)\,d\Sigma;$$

$$\text{Norm:} \quad \|f\|_{L^2(\Sigma)} = \left[\int_{\Sigma} f(s,t)^2\,d\Sigma\right]^{1/2}.$$

Let $L^2(\Sigma)\times L^2(\Omega) = V$. The inner product and norm of elements $[f_\Sigma f_\Omega]^T$, $[g_\Sigma g_\Omega]^T \in V$ are given by:

Inner product: $$(\mathbf{f},\mathbf{g})_V = \int_\Sigma f_\Sigma(s,t)g_\Sigma(s,t)\,d\Sigma + \int_\Omega f_\Omega(x)g_\Omega(x)\,dx\,;$$

Norm: $$\|\mathbf{f}\|_V = \left[\int_\Sigma f_\Sigma(s,t)^2\,d\Sigma + \int_\Omega f_\Omega^2(x)\,dx\right]^{1/2}.$$

Let $H^k(\Omega)$, $k = 1, 2, \ldots,$ be the kth-order (integer) Sobolev space of real functions whose squared powers and squared powers of all partial derivatives up to order k are integrable for the measure $dx = (dx_1, dx_2, \ldots, d_r)$. Define, for $\alpha \leq r$,

$$D^\alpha f(x)D^\alpha g(x) = \sum_{i_1,i_2,\ldots,i_\alpha=1}^{r} \frac{\partial^\alpha f(x)}{\partial x_{i_1}\partial x_{i_2}\ldots\partial x_{i_\alpha}}\,\frac{\partial^\alpha g(x)}{\partial x_{i_1}\partial x_{i_2}\ldots\partial x_{i_\alpha}}\,;$$

$$D^0 f D^0 g = f(x)g(x).$$

Then the inner product and norm of two elements $f, g \in H^k/(\Omega)$ are:

Inner product: $$(f,g)_{H^k(\Omega)} = \int_\Omega \sum_{\alpha=0}^{k} D^\alpha f(x)D^\alpha g(x)\,dx;$$

Norm: $$\|f\|_{H^k(\Omega)} = \left\{\int_\Omega \sum_{\alpha=0}^{k} D^\alpha f(x)D^\alpha f(x)\,dx\right\}^{1/2}.$$

(For $\alpha > r$ the notation is cumbersome. Suffice it to say that all partials up to order k are square integrable, and the inner product and norm are defined in a way consistent with the case $\alpha \leq r$.)

Let

$$H_0^k(\Omega) = \left\{f : f \in H^k(\Omega) \text{ and } \left.\frac{\partial^j f}{\partial v^j}\right|_\Gamma = 0, \quad 0 \leq j \leq k-1\right\},$$

where

$$\left.\frac{\partial f}{\partial v}\right|_\Gamma$$

is the derivative of f on Γ in the direction of the outward normal.

Finally, we shall be concerned with the space $L^2(0, T; P)$ of functions which for any time $t \in (0, T)$ are elements of H, and whose second powers are integrable with respect to the measure $dx\,dt$. Define the norm and inner product of two elements $f, g \in L^2(0, T; H)$:

$$\text{Inner product:} \quad (f, g)_{L^2(0,T;H)} = \int_0^T (f, g)_H \, dt ;$$

$$\text{Norm:} \quad \|f\|_{L^2(0,T;H)} = \left(\int_0^T \|f\|_H^2 \, dt \right)^{1/2} .$$

For clarity, we adopt the following notational conventions regarding f:

$f(x, t)$ is a point in R^1, where $x, t \in Q$;
$f(\cdot, t)$ is an element of the Hilbert space $H(\Omega)$;
$f(\cdot, \cdot)$ is an element of the Hilbert space $L^2(0, T; H)$;
$f(\cdot, \cdot; \alpha)$ is an element of the Hilbert space $L^2(0, T; H)$ parameterized by α, taken from a collection of functions $\{f_\alpha\}$ each an element of $L^2(0, T; H)$.

As we indicated, the class of functions $f(\cdot)$ which we shall consider are elements of some Hilbert space H. In general, the functions need not be continuous. We therefore define the derivative of these functions in the following way:

DEFINITION 2.1. Given a test function $\phi(\cdot) \in C^1(\Omega)$ with compact support in Ω; then, for $f(\cdot) \in L^2(\Omega)$, the mapping

$$\frac{\partial f}{\partial x_i} : \phi(\cdot) \rightarrow -\int_\Omega f \frac{\partial \phi}{\partial x^i} dx, \qquad i = 1, 2, \ldots, r \tag{2.1}$$

is called the distribution derivative or secant of the function f.

REMARK: The definition of $\partial f / \partial x^i$ is made in terms of a "supporting" function ϕ, as $\partial f / \partial x^i$ is a "weak" function, called a distribution or generalized function. We arrive at (2.1) in the following way: Integrate, by parts, the product

$$\frac{\partial f(x)}{\partial x_i} \phi(x)$$

over Ω to obtain, via Green's theorem;

$$\int_\Omega \frac{\partial f(x)}{\partial x_i} \phi(x) \, dx = -\int_\Omega f(x) \frac{\partial \phi(x)}{\partial x_i} dx .$$

For an example of the determination of a distribution derivative see Appendix 2.1.

It is convenient to express any function $f(.) \in H$ in terms of a countable, everywhere dense set of (known) "elementary" functions belonging to H.

Since we consider only separable Hilbert spaces, the existence of these functions is axiomatic. In particular, we consider as elementary functions the complete orthonormal system $\{w_i(x)\}_{i=1,2,\ldots}$ and state the following classical theorem:

THEOREM 2.1. In $L^2(\Omega)$ *space every complete orthonormal system is closed, and conversely.*

Thus, for any $f \in L^2(\Omega)$,

$$\lim_{m \to \infty} \left\| f - \sum_{i=1}^{m} c_i w_i \right\|^2_{L^2(\Omega)} \to 0,$$

where

$$c_i = \int_\Omega f(x) w_i(x)\, dx .$$

A special system of complete orthonormal functions which is used in the sequel are those generated by solutions to the classical *Sturm-Liouville* equation:

$$A[w] - \lambda \rho w = 0 \qquad \text{in } \Omega , \tag{2.2}$$

with any one of the three boundary conditions

$$\begin{aligned} &\text{i)}\ w(\Gamma) = 0, \\ &\text{ii)}\ \frac{\partial w(\Gamma)}{\partial v_A} = 0, \\ &\text{iii)}\ \frac{\partial w(\Gamma)}{\partial v_A} + \gamma(\Gamma) w(\Gamma) = 0, \qquad \sigma(\Gamma) > 0, \end{aligned} \tag{2.3}$$

with the hypothesis

$$A[w] = - \sum_{i,j=1}^{r} \frac{\partial}{\partial x_i} \left[a_{ij}(x) \frac{\partial w_j}{\partial x_j} \right] + a_0(x) w(x),$$

where

$a_0(x)$ and $a_{ij}(x)$ are bounded, measureable,

$a_{ij}(x) = a_{ji}(x)$ for all $x \in \Omega$,

$a_0(x) \geq \alpha > 0$ almost every where in $L^2(\Omega)$,

$\sum_{i,j=1}^{r} a_{ij}(x) \xi_i \xi_j \geq \alpha(\xi_1^2 + \xi_2^2 + \ldots + \xi_r^2)$ almost everywhere in Ω for all $\xi = (\xi_1 \xi_2 \cdots . \xi_r)$ in R^r,

$f(x) \geq 0$ almost everywhere in Ω.

In summary, we give the classical theorem (see, for example, [15])

THEOREM 2.2. *Solutions to* (2.2) *and any one of* (2.3) *constitute a complete orthonormal system in* $L^2(\Omega)$. *These solutions* $w_i(x)$, $x \in \Omega$, *are called the eigenfunctions of the Sturm-Liouville problem or, more simply, eigenfunctions.*

Of special interest are the eigenfunctions of

$$\frac{\partial^2 w}{\partial x^2} + \lambda w = 0 \qquad \text{on } (0,1),$$

$$w(0) = 0,$$

$$w(1) = 0,$$

namely,

$$\{\sqrt{2}\sin\sqrt{\lambda_i}x\}_{i=1,2,\ldots}, \qquad \lambda_i = (i\Pi)^2$$

A.2 SOLUTION PROPERTIES OF LINEAR PARABOLIC AND SECOND-ORDER HYPERBOLIC PARTIAL DIFFERENTIAL EQUATIONS

We give the following lemmas which, under appropriate hypothesis (given), assert the existence and uniqueness of solutions to two classes of linear partial differential equations. These equations, along with the pertinent hypothesis, are introduced formally here. A more natural introduction to these equations is given in Chapter 3.

A.2.1 Linear Parabolic Equations

$$\frac{\partial y(x,t;\mathbf{u})}{\partial t} + A[y(x,t;\mathbf{u})] = f(x,t), \qquad x,t \in Q. \tag{2.4}$$

Initial condition:

$$y(x,0) = u_2(x), \qquad x \in \Omega. \tag{2.5}$$

Boundary condition—one of:

(I)
$$y(s,t) = u_1(s,t), \qquad s,t \in \Sigma, \tag{2.6}$$

(II)
$$\frac{\partial y(s,t)}{\partial \nu_A} = u_1(s,t), \qquad s,t \in \Sigma, \tag{2.7}$$

(III)
$$\frac{\partial y(s,t)}{\partial \nu_A} + \beta(s,t)\,y(s,t) = u_1(s,t), \qquad s,t \in \Sigma, \tag{2.8}$$

where

$$\mathbf{u} = [u_1(s,t)u_2(x)]^T .$$

REMARK: It is possible to consider boundary conditions which are combinations of (I), (II), and (III) in the following sense: Γ is partitioned in to $\Gamma_1\Gamma_2\ldots\Gamma_k$ such that $\bigcup_k \Gamma_k = \Gamma$. Then, on each of the sets $\Sigma_i = \Gamma_i \times (0,T]$, one of the conditions (I), (II), or (III) holds. Such considerations introduce no new technical problems and are not considered.

(i) The elliptic operator $A[\varphi(x,t)]$ may now have time variable coefficients:

$$A[\varphi(x,t)] = -\sum_{i,j=1}^{r} \frac{\partial}{\partial x_i}\left[a_{ij}(x,t)\frac{\partial}{\partial x_j}\varphi(x,t)\right] + a_0(x,t)\varphi(x,t),$$

$$a_0(x,t) \geq \alpha > 0 \qquad \text{almost everywhere in } \Omega \text{ for all } t\in(0,T];$$

$$\sum_{i,j=1}^{r} a_{ij}(x,t)\xi_i\xi_j \geq \alpha(\xi_1{}^2 + \xi_2{}^2 + \ldots + \xi_r{}^2)$$

almost everywhere in Ω for all $t\in(0,T]$;

$a_0(x,t)$, $a_{ij}(x,t)$ bounded, measurable, $\quad$ i, j $=1,2\ldots r$;

In addition,

$$\beta(s,t)\geq 0 \qquad \text{for all } s,\ t\in\Sigma .$$

(ii) $f(.,.)L^2(Q)$.
(iii) $u_2(\cdot)\in L^2(\Omega)$.
(iv) $u_1(.)\in L^2(\Sigma)$.

Nomenclature. We shall refer to equations (2.4), (2.5), and (2.6) as *system (I)*, (2.4), (2.5), and (2.7) as *system (II)*, and (2.4), (2.5), and (2.8) as *system (III)*.

Notation.

$$\frac{\partial y}{\partial \nu_A} = \sum_{i,j=1}^{r} a_{ij}(s,t)\frac{\partial y}{\partial x_j}\cos(\mathbf{n},x_i)_{L^2(R^r)},$$

where $\mathbf{n}$ = vector *outward* normal to the curve Γ.

LEMMA 2.1. *If a solution exists for system (I), system (II), or system (III) such that $y(\cdot,\cdot)\in L^2(0,T;H^1)$, then in each case that solution is unique.*

A proof is given in Appendix 2.5.

LEMMA 2.2. (*Lions-Magnenes*). *Under the given hypothesis a solution to system (I) exists in $L^2(Q)$ and is unique. Note that*

$$\frac{\partial y(.,t;\mathbf{u})}{\partial x_i} \notin L^2(\Omega), \qquad i = 1,2,\ldots,r.$$

LEMMA 2.3. *Under the given hypothesis a solution to system (II) exists such that*

$$\int_0^T \|y(t)\|_{H^1(\Omega)}^2 \, dt < \infty .$$

LEMMA 2.4. *Under the given hypothesis a solution to system (III) exists such that*

$$\int_0^T \|y(t)\|_{H^1(\Omega)}^2 \, dt < \infty .$$

REMARK. The proof of Lemmas 2.2, 2.3, and 2.4 may be found in [1].

Thus we have the existence and uniqueness of solutions to systems (I), (II), and (III) (of appropriate class).

A.2.2 Linear, Second-Order Hyperbolic Equations

$$\frac{\partial^2 y(x,t;\mathbf{u})}{\partial t^2} + A[y(x,t;\mathbf{u})] = f(x,t), \qquad x,t \in Q . \tag{2.9}$$

Initial conditions:

$$y(x,0) = u_2(x), \qquad x \in \Omega , \tag{2.10}$$

$$\frac{dy(x,0)}{dt} = u_3(x), \qquad x \in \Omega . \tag{2.11}$$

Boundary conditions—one of:

(I) $$y(s,t) = u_1(s,t), \qquad s,t \in \Sigma , \tag{2.12}$$

(II) $$\frac{\partial y(s,t)}{\partial \nu_A} = u_1(s,t), \qquad s,t \in \Sigma , \tag{2.13}$$

(III) $$\frac{\partial y(s,t)}{\partial \nu_A} + (s,t)\, y(s,t) = u_1(s,t), \qquad s,t \in \Sigma . \tag{2.14}$$

Hypothesis. We augment the hypothesis of Section A.2.1 with the following:

$$a_{ij}(x,t) = a_{ji}(x,t) \qquad (i,j = 1, 2, \ldots, r) \qquad \text{for all } x,t \in Q .$$

$$a_{ij}(x,t) \text{ is } C^1(0,T).$$

Nomenclature. Equations (2.9), (2.10), and (2.11) with the boundary conditions (I), (II), or (III) are called system (I), (II), and (III), respectively.

For systems (I), (II), and (III) we state the following three lemmas, analogous to lemmas 2.2, 2.3, and 2.4 of Section A.2.1:

LEMMA 2.5. *Under the augmented hypothesis, a solution to system (I) exists in* $L^2(0,T;H)$ *and is unique.*

LEMMA 2.6. *Under the augmented hypothesis a solution to system (II) exists in* $L^2(0,T;H^1)$ *and is unique.*

LEMMA 2.7. *Under the augmented hypothesis a solution to system (III) exists in* $L^2(0,T;H^1)$ *and is unique.*

Thus the solutions to each of systems (I), (II), and (III) of Sections A.2.1 and A.2.2 exist (in the appropriate space) and, moreover, are unique. Another property of these systems is that they are "well set" in the sense of Hadamard, a feature which is summarized in the following Lemma:

LEMMA 2.8. *The solutions of each of the systems (I), (II), and (III) of Sections* A.2.1 *and* A.2.2 *vary continuously with the initial and boundary data. That is, for any* $\varepsilon > 0$, *there exists a* δ *such that*

$$\|\mathbf{u}-\mathbf{v}\|_V < \delta, \qquad \mathbf{u}, \mathbf{v} \in V \tag{2.15}$$

implies that

$$\|y(x,t;\mathbf{u}) - y(x,t;\mathbf{v})\|_{L^2(Q)} < \varepsilon . \tag{2.16}$$

We give an independent proof in Appendix 2.6.

Lemma 2.8 establishes the stability of the solutions to systems (I), (II), and (III) of Sections A.2.1 and A.2.2 (stability in the sense that "small" excursions from some nominal boundary and initial data give rise to "small" trajectory variations of the response).

Attention is now directed toward a technique which uses the result of Theorem 2.1 to obtain approximate solutions to the systems (I), (II), and (III).

A.3 THE GALERKIN APPROXIMATION SCHEME

As we indicated in Section A.1, it is frequently convenient to describe a function from a given class in terms of a set of elementary functions which form a basis in that class. Here, in the context of solving a given type of partial differential equation, we are concerned with approximating the solution of that equation, which is known to belong to a certain class of functions.

We are given, for example, that $y(\cdot, t; \mathbf{u})$ is the solution of a certain differential equation. Moreover, $y(\cdot, t; \mathbf{u}) \in H$, some Hilbert space. If $\{w_i\}_{i=1,2,\dots}$ are a basis in H, then we seek approximations to $y(\cdot, t; \mathbf{u})$ of the form

$$y_m(x, t; \mathbf{u}) = \sum_{i=1}^{m} y_i(t; \mathbf{u}) w_i(x) . \tag{2.17}$$

We shall consider the cases where $\{w_i\}_{i=1,2,\dots}$ are

(A) the Normalized eigenfunctions in H, and

(B) the cubic splines, for Ω a segment of R^1.

REMARK. (i) The problem of obtaining the eigenfunctions to the Sturm-Liouville equation (Section A.1) when $\Omega \bigcup \Gamma$ is "irregular," that is, when $\Omega \bigcup \Gamma$ is not a simple geometric figure, is complex. We accept this restriction, and proceed under its shadow.

(ii) No approximations arising from the common "differencing" methods are attempted, except in the sense that (B) is sophisticated differencing technique.

To illustrate the Galerkin method, consider its application to system (I) of Section A.2.1.

In preparation for the application of this approximation technique to a system with nonhomogeneous boundary conditions, it is customary first to employ an affine trnasformation of the dependent variables so as to convert the original system into two systems as follows:

Define

$$y(x, t) = u(x, t) + w(x, t) . \tag{2.18}$$

According to (2.4),

$$\frac{\partial u(x,t)}{\partial t} + A[u(x,t)] = f(x,t) - \left\{\frac{\partial w(x,t)}{\partial t} + A[w(x,t)]\right\}. \tag{2.19}$$

As usual, a function $w(x,t)$ is constructed, satisfying

$$w(s, t) = u_1(s, t), \qquad s, t \in \Sigma .$$

Now we observe that the hypothesis on $u(\Sigma)$ is that $u(.) \in L^2(\Sigma)$. Evidently, difficulties may arise when the necessary derivatives are taken on the function $w(x, t)$, $x, t \in Q$, since the RHS of (2.19) must be $L^2(Q)$. These difficulties are skirted by proceeding in a novel way, proposed by Lions [16].

Define
Ψ is a function with the following properties:

$$\frac{\partial \Psi(x,t)}{\partial t} + A[\Psi(x,t)] = \rho(x,t), \qquad \rho(\cdot,\cdot) \in L^2(Q), \tag{2.20}$$

$$\Psi(s,t) = 0, \tag{2.21}$$

$$\Psi(x,T) = 0. \tag{2.22}$$

Multiply (2.4) by $\Psi(x,t)$ and integrate over Q to obtain:

$$\int_Q y(x,t)\rho(x,t)\,dx\,dt - \int_\Omega u_2(x)\,\Psi(x,0)\,dx$$
$$+ \int_\Sigma u_1(\Sigma)\frac{\partial \Psi(\Sigma)}{\partial \nu_A}\,d\Sigma = \int_Q f(x,t)\,\Psi(x,t)\,dx\,dt. \tag{2.23}$$

Equation (2.23) is a linear integral equation of the first type and is equivalent to the system (I) of Section A.2.1.

The application of the Galerkin technique to (2.23) is straightforward provided $\Psi(x,t)$ is defined.

REMARK. The following developments are valid for type (A) approximations. Since type (B) approximations are essentially different, the corresponding approximation technique is considered in Appendix 2.3.

Define

$$\Psi(x,t) = g(t)w_j(x), \qquad j \text{ fixed but arbitrary}, \tag{2.24}$$

$$g(\cdot) \in C^1(0,T), \qquad g(T) = 0. \tag{2.25}$$

$$u_{1i}(t) = \int_\Gamma \frac{\partial w_i(\Gamma)}{\partial \nu_{A^*}} u_1(\Sigma)\,d\Gamma, \tag{2.26}$$

$$y_m(x,t) = \sum_{i=1}^{m} y_i(t)w_i(x). \tag{2.27}$$

Evidently,

$$\rho(x,t) = \left[-\frac{dg(t)}{dt} + \lambda_j g(t)\right] w_j(x), \qquad \rho(\cdot,\cdot) \in L^2(Q),$$

$$\Psi(s,t) = 0, \qquad s,t \in \Sigma,$$

$$\Psi(x,T) = 0, \qquad x \in \Omega,$$

$$\lim_{m\to\infty} y_m(\cdot,t) \to y(\cdot,t) \in L^2(\Omega), \qquad \text{for each } t \in (0,T].$$

Using definitions (2.24), (2.25), (2.26), and (2.27) and the properties of $\{w_i\}_{i=1,2,\ldots}$ in (2.23), there results:

$$\int_0^T y_j(t)\left\{-\frac{dg}{dt}+\lambda_j g(t)\right\}dt-g(0)u_{2j}+\int_0^T g(t)u_{1j}(t)\,dt$$

$$=\int_0^T f_j(t)g(t)\,dt, \tag{2.28}$$

where we used

$$f(x,t)=\lim_{m\to\infty}\sum_{i=1}^{m} f_i(t)w_i(x), \qquad f_i(t)=\int_\Omega f(x,t)w_i(x)\,dx,$$

$$u_2(x)=\lim_{m\to\infty}\sum_{i=1}^{m} u_{2i}w_i(x), \qquad u_{2i}=\int_\Omega u_2(x)w_i(x)\,dx.$$

Integrating (2.28) by parts and using the properties of $g(t)$, we obtain the following equation for $y_j(t)$, $j=1,2,\ldots,m$):

$$\frac{dy_j(t)}{dt}+\lambda_j(t)y_j(t)=f_j(t)-u_{1j}(t),$$

$$y_j(0)=u_{2j}, \qquad j=1,2,\ldots,m. \tag{2.29}$$

We call (2.27) the Galerkin approximation to system (I), with $y_j(t)$ defined by (2.29), $j=1,2,3,\ldots,m$.

It is possible to obtain an estimate of the error of approximation. Indeed, if we define:

$$M=\sup_i \sup_{t\in(0,T]}\{|f_i(.)-u_{1i}(\cdot)|\}. \tag{2.30}$$

Then it is shown in Appendix 2.2 that:

$$E_m(y_m)=\|y(.,t)-\sum_{i=1}^{m} y_i(t)w_i(\cdot)\|_{L^2(\Omega)}\le\left[\frac{M^2}{3\Pi^4}\right]^{1/2}\left(\frac{1}{m}\right)^{3/2},$$

that is, the error is of the order $(1/m)^{3/2}$.

REMARK. The error estimate obtained for the Galerkin approximation using the eigenfunctions as a basis indicates that the convergence may be slow. (It is of course possible that there is no error after m terms). Furthermore, the convergence of the derivatives of y_m to those of y is slow. A more strict hypothesis on the function $y(\cdot,t)$ being approximated, for example, $y(\cdot,t)\in C^4(\Omega)$ allows the adoption of splines as a "basis" and some striking error estimates result. We consider the spline approximation in Appendix 2.3 and point out an unfortunate drawback which hinders their application *in the context of this study.*

We consider next some fundamental results pertaining to the characterization of extremals of quadratic functionals on H.

A.4 LINEAR, BILINEAR, AND QUADRATIC FUNCTIONALS ON A HILBERT SPACE

Let **u**, **v**, and **w** be elements of the Hilbert space V defined in Section A.1. Define the following:

(i) $\mathrm{I}(\mathbf{v})$ is a continuous linear functional on V. That is,

$$|\mathrm{I}(\mathbf{v})| \leq M\|\mathbf{v}\|_V , \qquad M < \infty ;$$

for $\alpha, \beta \in R^1$,

$$\mathrm{I}(\alpha\mathbf{u}+\beta\mathbf{v}) = \alpha\mathrm{I}(\mathbf{u})+\beta\mathrm{I}(\mathbf{v}) .$$

(ii) $a(\mathbf{u}, \mathbf{v})$ is a continuous bilinear functional on V. That is,

$$|a(\mathbf{u}, \mathbf{v})| \leq N[\mathbf{u}\|_V \|\mathbf{v}\|_V ; \qquad N < \infty ;$$

for $\alpha, \beta \in R^1$,

$$a(\alpha\mathbf{u}+\beta\mathbf{v}, \mathbf{w}) = \alpha a(\mathbf{u}, \mathbf{w})+\beta a(\mathbf{v}, \mathbf{w})$$

and

$$a(\mathbf{u}, \alpha\mathbf{v}+\beta\mathbf{w}) = a\alpha(\mathbf{u}, \mathbf{v}) +\beta a(\mathbf{u}, \mathbf{w}) .$$

(iii) K is a closed convex set in V.
(iv) $J(\mathbf{v})$ is a quadratic functional on V with the property

$$J(\mathbf{v}) = a(\mathbf{v}, \mathbf{v})-2\mathrm{I}(\mathbf{v}) .$$

LEMMA 2.9. *$J(\mathbf{v})$ is a convex functional on V: for $\mathbf{u}, \mathbf{v} \in V$ and $0 \leq \lambda \leq 1$,*

$$J(\mathbf{u}+(1-\lambda)\mathbf{v}) \leq \lambda J(\mathbf{u})+(1-\lambda)J(\mathbf{v}) .$$

The proof of Lemma 2.9 is well known and is given in Appendix 2.4.

We give a theorem which asserts the existence and uniqueness of an element $\mathbf{u} \in K$ with the property:

$$J(\mathbf{u}) = \inf_{\mathbf{v} \in V} J(\mathbf{v}) .$$

THEOREM 2.3. *There exists a $\mathbf{u} \in K$, unique such that*

$$J(\mathbf{u}) = \inf_{v \in K} J(\mathbf{v}), \qquad J(\mathbf{u}) \leq J(\mathbf{v}) \quad \text{for all } \mathbf{v} \in K.$$

PROOF. See Appendix 2.7.

The following two theorems which characterize the $\mathbf{u} \in K$ for which

$$J(\mathbf{u}) = \inf_{\mathbf{v} \in K} J(\mathbf{v})$$

are due to Lions and Stampaccia [14]. The proofs are included here because of the importance of these theorems.

THEOREM 2.4. *The unique element* $\mathbf{u} \in K$ *for which*

$$J(\mathbf{u}) = \inf_{\mathbf{v} \in K} J(\mathbf{v})$$

is characterized by:

$$a(\mathbf{u}, \mathbf{v}-\mathbf{u}) \geq \mathrm{I}(\mathbf{v}-\mathbf{u}) \qquad \textit{for all } \mathbf{v} \in K. \tag{2.31}$$

PROOF (i) *Necessity.*

$$J(\mathbf{u}) \leq J(\mathbf{w}) \qquad \text{for all } \mathbf{w} \in K. \tag{2.32}$$

Let $\mathbf{w} = (1-\lambda)\mathbf{u} + \lambda\mathbf{v}$, $0 \leq \lambda \leq 1$. *Then*

$$J(\mathbf{u}) \leq J((1-\lambda)\mathbf{u} + \lambda\mathbf{v}). \tag{2.33}$$

We have that

$$J((1-\lambda)\mathbf{u}+\lambda\mathbf{v}) = J(\mathbf{u}+\lambda(\mathbf{v}-\mathbf{u})) = J(\mathbf{u}) + 2a(\mathbf{u}, \mathbf{v}-\mathbf{u})$$
$$+\lambda^2 a(\mathbf{v}-\mathbf{u}, \mathbf{v}-\mathbf{u}) - 2\mathrm{I}(\mathbf{v}-\mathbf{u}).$$

Taking note of (2.33), then

$$2a(\mathbf{u}, \mathbf{v}-\mathbf{u}) - 2\mathrm{I}(\mathbf{v}-\mathbf{u}) + \lambda a(\mathbf{v}-\mathbf{u}, \mathbf{v}-\mathbf{u}) \geq 0, \qquad 0 \leq \lambda \leq 1 \tag{2.34}$$

Set $\lambda = 0$ in (2.34) and obtain the desired result given by (2.31).

(ii) *Sufficiency.*

$$J(\mathbf{v}) - J(\mathbf{u}) = a(\mathbf{v}, \mathbf{v}) - a(\mathbf{u}, \mathbf{u}) - 2\mathrm{I}(\mathbf{v}-\mathbf{u}).$$

Use the result that

$$a(\mathbf{u}, \mathbf{v}-\mathbf{u}) \geq \mathrm{I}(\mathbf{v}-\mathbf{u})$$

and obtain

$$J(\mathbf{v}) - J(\mathbf{u}) \geq a(\mathbf{v}, \mathbf{v}) - a(\mathbf{u}, \mathbf{u}) - 2a(\mathbf{u}, \mathbf{v}-\mathbf{u}) = a(\mathbf{v}-\mathbf{u}, \mathbf{v}-\mathbf{u})$$
$$\geq \alpha \|\mathbf{v}-\mathbf{u}\|_V^2 \geq 0, \qquad \mathbf{v} \in K$$

THEOREM 2.5. *For* $K = V$, *then Theorem* 2.4 *implies that the unique* $\mathbf{u} \in V$ *for which* $J(\mathbf{u}) = \inf_{\mathbf{v} \in V} J(\mathbf{v})$ *is characterized by*

$$a(\mathbf{u}, \mathbf{v}) - \mathrm{I}(\mathbf{v}) = 0. \tag{2.35}$$

PROOF. Let $\mathbf{v} = \mathbf{u} + \mathbf{e}$, $\quad \mathbf{e}, \mathbf{v} \in V$.
Put this choice of $\mathbf{v}$ into (2.31,) and obtain

$$\pm a(\mathbf{u}, \mathbf{e}) \geq \pm \mathrm{I}(\mathbf{e})$$
$$\Rightarrow$$
$$a(\mathbf{u}, \mathbf{e}) = \mathrm{I}(\mathbf{e}) \qquad \text{for all } \mathbf{e} \in V.$$

Theorems 2.4 and 2.5 provide the theoretical basis for a variational approach to the identification problem to be introduced in Chapter 3 and developed in Chapters 4 and 5. We remark that theorems 2.4 and 2.5 are a maximum (or, more properly, a minimum) principle which is at once more general and less general than the celebrated maximum principle of Pontryagin [17]. It is more general in the sense that, as we will show, it characterizes extremals to systems whose evolution processes are distributed. It is less general because of the hypothesis on $J(\mathbf{v})$ and also because it is applicable to linear evolution processes only.

It is possible to obtain other characterizations of the optimal $\mathbf{u}$ for other special types of closed and convex K, as for example when K is a cone with vertex at the origin. However, the identification problems posed in Chapter 3 do not need such hypothesis.

It should be remarked that Theorems 2.4 and 2.5 have wider applicability than that which is given in this study. Indeed, Lions has used these results to obtain the first rigorous treatment of a class of distributed optimal control problems [6]. Applications are also immediate in the treatment of some *classical* problems in the mathematical physics, such as the "maximum principle" for elliptic partial differential equations.

CHAPTER 3

FORMAL DEFINITION OF THE STATE IDENTIFICATION PROBLEM

A.0 INTRODUCTION

The state identification problem was informally introduced in Chapter 1. Here, in Section A.1, the systems whose states are to be identified are presented along with a precise statement of the criteria of identification (Section A.2). Section A.3 deals with a special relation which exists between two identification criteria.

A.1 DEFINITION OF THE CLASS OF SYSTEMS CONSIDERED

The abstract system from which an input-output record has been obtained is assumed to have associated with it the state-determined character. That is, the output at some time $t > t_0$ is uniquely determined by knowledge of the input record over $t_0 \leqslant \tau \leqslant t$ and the evolution of the *state* of the system, starting from a known value at $t = t_0$. The state evolution process is described mathematically by a partial differential equation. Conventionally, the system is classified by the type of equation satisfied by the associated state evolution process. We shall investigate the state identification of linear parabolic and second-order hyperbolic systems.

The domain of dependence of these partial differential equations is customarily space and time. At any time $t > t_0$, $y(\cdot, t)$, the system state, is therefore a function of the space variable(s). Generally, an infinite number of elementary or basis functions would be necessary to describe the state at time $t > t_0$. Thus the state space for the systems under consideration are "infinite dimensional." We assume without much loss of generality that the state equations are scalar.

A.1.1 Systems Classification

The classification of parabolic or (second-order) hyperbolic equations arises when the following general linear partial differential equation with real coefficients is transformed to a canonical form:

$$\sum_{j=1}^{r+1} \sum_{j=1}^{r+1} a_{ij}(\mathbf{x}) \frac{\partial^2 u(\mathbf{x})}{\partial x_i \partial x_j} + \sum_{i=1}^{r+1} b_i(\mathbf{x}) \frac{\partial u(\mathbf{x})}{\partial x_i} + c(\mathbf{x})u(\mathbf{x}) + f(\mathbf{x}). \qquad (3.1)$$

Equation (3.1) is reducible to *one* of the following three canonical forms:†

$$\frac{\partial^2 u(\mathbf{x})}{\partial x_1{}^2} + \frac{\partial^2 u(\mathbf{x})}{\partial x_2{}^2} + \dots + \frac{\partial^2 u(\mathbf{x})}{\partial x_{r+1}^2} + g(\mathbf{x}) = 0 \qquad \text{(elliptical type)} \qquad (3.2)$$

$$\frac{\partial^2 u(\mathbf{x})}{\partial x_1{}^2} = \sum_{i=2}^{r+1} \frac{\partial^2 u}{\partial x_i{}^2} + g(\mathbf{x}) \qquad \text{(hyperbolic type)} \qquad (3.3)$$

$$\sum_{i=1}^{r+1-m} (\pm u_{x_i x_i}) + g(\mathbf{x}) = 0, \qquad 0 < m < r+1 \qquad \text{(parabolic type)} \qquad (3.4)$$

For details on the method of classification, see, for example, [18]. This classification is given here to serve as a bridge between the equations which arise in practice and those canonical forms exposed in Chapter 2. We remark here that much of the literature on distributed systems contains canonical forms far more general than those considered in this monograph. This generality is, however, illusory, since the theoretical results are weakened by assumptions, explicit or implicit, not made here. For example, it is not necessary for us to assume existence of solutions of a certain class. Admittedly, assumptions of this nature are frequently substantiated and play no adverse role in algorithmic solutions, but we strive here for some rigor. Having solved the more restrictive classes of problems, points of departure suggest themselves. Thus we consider the linear parabolic and second-order hyperbolic equations of evolution.

A.1.2 Examples of Parabolic and Hyperbolic Systems

The list of physical systems which may be categorized as either parabolic or hyperbolic is long. Consider the following salient examples.

(a) *Parabolic:* Temperature $y(x, t)$ at a point $x \in \Omega$ and $t \in (0, T)$ of a medium which is exchanging heat in a predominantly diffusive manner with its environment, which at that point in space and that time is at a temperature $f(x, t)$:

$$\frac{\partial y(x, t)}{\partial t} - k \frac{\partial^2 y(x, t)}{\partial x^2} = f(x, t). \qquad (3.5)$$

† We make no assumptions here on the coefficients $a_{ij}(\mathbf{x})$, $b_i(\mathbf{x})$, and $c(\mathbf{x})$ nor on the differentiability of $u(\mathbf{x})$.

The boundary conditions of the system represented by (3.5) are exhausted by three considerations:

(i) No insulation.
(ii) Full insulation.
(iii) Partial insulation.

These three physical orientations are mathematically represented by conditions (I), (II), (III) (respectively) of Section A.2.1 of Chapter 2.

REMARKS: There are systems described by a set of vector equations which are hybrids: parabolic in one dependent variable and hyperbolic in another. Since we consider the scalar differential equation exclusively, such systems are bypassed (though not because of any inherent difficulty).

The "transport" term $\partial y/\partial x$ occurring in the description of many heat exchange problems can be avoided by a linear transformation leading to the form given in (3.5).

(b) *Hyperbolic*: Vibrations in an elastic medium. Consider for example the case of longitudinal displacement $y(x,t)$ at a point x in an elastic rod, at a certain time t, experiencing a dynamic axial load applied at the ends:

$$\frac{\partial^2 y(x,t)}{\partial t^2} - \frac{\rho}{EA}\frac{\partial^2 y(x,t)}{\partial x^2} = 0. \tag{3.6}$$

The load is applied in one of three ways:

(i) Direct axial loading.
(ii) Transverse bending.
(iii) Loading through an elastic support.

Cases (i), (ii), and (iii) correspond to (2.12), (2.13), and (2.14).

The examples given are classical. The question of identification of such systems may seem of academic importance until it is realized that more modern systems can be cast into the framework of the two illustrative classes. For example, a little reflection casts the problem of vibrations in a slender airframe into the perspective of example (b), appropriately modified. Also seismic signals, wherein the propagation of waves (in an unbounded elastic region) is used to detect anomalous objects, form a large class of important problems. These problems arise, for example, in nondestructive testing, geophysical exploration, and naval surveillance of underwater objects. An exact definition of identification is now undertaken.

REMARKS: It is possible to consider combinations of (i), (ii), or (iii) on the boundary; for example, condition (i) at one end of the rod and (ii) at the other. In the more general setting of multidimensional spatial domains such combinations are of rich variety.

All our results are valid for any such consistent combinations.

A.2 STATE IDENTIFICATION

The concept of state identification can be introduced by consideration of the following problem.

Suppose that a real physical system S has been evolving from some indeterminate time past and that the intitial data with whieh the evolution began are known inexactly. Then, for the distributed systems under study, the state at some point M in the space-time domain is parameterized by the initial data and the values of the forcing on the boundary and in the interior. Measurements I and O are allowed in the input and output spaces, respectively, of the system S. The question arises, in what way can these measurements be utilized so that

(a) a "refinement" R of these inexact measurements is accomplished, consistent with

(b) an extension of the information contained in the results, accomplished via knowledge of the state evolution process.

The resolution of this question constitutes a solution to the state identification problem. The statement that a "refinement" of the measurement process is sought implies that a criterion of satisfaction is adopted. The criterion adopted throughout is that of "least squares," which is defined presently. The identification problem has the following,

A.2.1 Mathematical Description

System S. The systems are as defined in Sections A.2.1 and A.2.2 of Chapter 2.

Input Measurements I. The inexactly known inputs occur on the boundary of the system S:

$$z_1(\cdot) = u_1(\cdot)+\varepsilon_1(\cdot) \quad \text{on } \Sigma, \qquad z_1(\cdot)\in L^2(\Sigma)\,. \tag{3.7}$$

$$z_2(\cdot) = u_2(\cdot)+\varepsilon_2(\cdot) \quad \text{in } \Omega \qquad z_2(\cdot)\in L^2(\Omega)\,. \tag{3.8}$$

The input I can be considered as a vector:

$$\mathbf{z} = \{z_1(\Sigma)|z_2(x)\}^T = \{u_1(\Sigma)|u_2(x)\}^T+\{\varepsilon_1(\Sigma)|\varepsilon_2(x)\}^T\,,$$

$$\mathbf{z}\in L^2(\Sigma)\times L^2(\Omega); \qquad \mathbf{u}\in L^2(\Sigma)\times L^2(\Omega)\,.$$

That is,

$$\int_{\Sigma} z_1{}^2(\Sigma)\,d\Sigma + \int_{\Omega} z_2(x)^2\,dx < \infty, \qquad \int_{\Sigma} u_1(\Sigma)^2\,d\Sigma + \int_{\Omega} u_2(x)^2\,dx < \infty .$$

REMARKS: We assume that forcing of the states in the interior of the spatial domain, if present, is known with precision.

The inexactly known initial condition $u_2(\cdot)$ is considered an input to the system S.

Output Measurements O: In general, the output measurements on the system state $y(x, t; \mathbf{u})$ are given by:

$$z(x, t) = R(x, t)[y(., \cdot; \mathbf{u})] + \varepsilon(x, t), \qquad x\varepsilon\Omega, \quad t\varepsilon(0, T], \tag{3.9}$$

where

$$R(x, t)[\cdot]\colon L^2(Q) \to L^2(Q) \qquad \text{(Linear)} .$$

In particular, we consider the following three forms of (3.9) in detail:

$$\text{(I)}\quad R(x, t)\,[y(\cdot, \cdot; \mathbf{u})] = \int_{\Omega} y(\xi, t)\,\delta(\xi - x)\,d\xi = y(x, t),$$

$$z(x, t) = y(x, t) + \varepsilon(x, t), \qquad z(\cdot, \cdot) \in L^2(Q). \tag{3.10}$$

$$\text{(II)}\quad R(x, t)\,[y(\cdot, \cdot; \mathbf{u})] = \int_{\Omega} \sum_{i=1}^{v} y(\xi, t)\,\delta(\xi - x^i)\,d\xi = \sum_{i=1}^{v} y(x^i, t), \tag{3.11}$$

$$z(x^i, t) = y(x^i, t) + \varepsilon(x^i, t), \qquad z(x^i, \cdot) \in L^2((0, T]),$$

where x^i is a point in the spatial domain in Ω.

$$\text{(III)}\quad R(x, t)[y(\cdot, \cdot; \mathbf{u})] = 0, \tag{3.12}$$

$$z(x, t) = 0 \quad \text{everywhere in } Q.$$

In (3.7), (3.8) and (3.9) the error processes $\varepsilon_1(\Sigma)$, $\varepsilon_2(x)$, and $\varepsilon(x, t)$ are assumed to be unknown (except with respect to the fact that they are square integrable) In Section A.3 some knowledge of a statistical nature is assumed for these processes.

Data Refinement R. The dependence of the ıesponse $y(x, t)$ of the system S on the boundary and initial data $\mathbf{u}$ has already been mentioned. In order to display this dependence explicitly, write

$$y(x, t) = y(x, t; \mathbf{u})$$

Define

$\mathbf{v}$ = an arbitrary estimate of the "true state of nature," $\mathbf{u}^*$;
$\mathbf{u}$ = a refined estimate of the true state of nature $\mathbf{u}^*$ given measurements z, z_1, and z_2 and the equation of evolution of the system S.

The refinement R is chosen to be in the least squares sense: $\mathbf{u}$ minimizes the quadratic functional

$$J(\mathbf{v}) = \int_Q [\mathbf{R}[y(x,t;\mathbf{v})] - z(x,t)]^2 dx\,dt + \int_\Sigma [v_1(\Sigma) - z_1(\Sigma)]^2 d\Sigma + \int_\Omega [v_2(x) - z_2(x)]^2 dx. \tag{3.13}$$

In (3.13), $y(x, t; \mathbf{v})$ evolves according to the definition of S. In order to stress some important properties of the refinement chosen, we generalize the statement of (3.13) in the following way:

$$J(\mathbf{v}) = a(\mathbf{v},\mathbf{v}) - 2\mathrm{I}(\mathbf{v}) + k, \tag{3.14}$$

where $a(\mathbf{v},\mathbf{v})$ and $\mathrm{I}(\mathbf{v})$ are endowed with the properties given in Chapter 2, Section A.4. These properties are restated here:

(i) $a(\mathbf{v},\mathbf{w})$ is a bilinear, continuous form on the Hilbert space $L^2(\Sigma) \times L^2(\Omega)$.
(ii) There exists an $\alpha > 0$ such that

$$a(\mathbf{v},\mathbf{v}) \geq \alpha \|\mathbf{v}\|_{L^2(\Sigma)\times L^2(\Omega)}.$$

(iii) $a(\mathbf{u},\mathbf{v}) = a(\mathbf{v},\mathbf{u})$.
(iv) $\mathrm{I}(\mathbf{v})$ is a continuous linear form on $L^2(\Sigma) \times L^2(\Omega)$.

It is essential that the functional $J(\mathbf{v})$ given by (3.13) have all the properties of the quadratic form (3.14). The theorems given in Chapter 2 characterizing extremals of a quadratic functional will furnish a solution to the identification problem given. It is essential that the corresponding hypothesis be compatible. We now establish that for the system S and any of the measurement processes (I), (II), or (III), (3.13) can be phrased as in (3.14).

THEOREM 3.1. *Equations* (3.13) *and* (3.14) *are equivalent if* $a(\mathbf{v},\mathbf{v})$ *and* $\mathrm{I}(\mathbf{v})$ *are defined by* (3.15), (3.16), *and* (3.17). *Moreover,* $a(\mathbf{v},\mathbf{v})$ *and* $\mathrm{I}(\mathbf{v})$ *obey the hypothesis of Theorem* 2.5:

$$a(\mathbf{v},\mathbf{v}) = \int_Q \{L[y(x,t;\mathbf{v}) - y(x,t;\mathbf{0})]\}^2 dx\,dt + \int_\Sigma v_1(\Sigma)^2 d + \int_\Omega v_2(x)^2 dx. \tag{3.15}$$

$$\mathrm{I}(\mathbf{v}) = -\int_Q R[y(x,t;\mathbf{v}) - y(x,t;\mathbf{0})]\{R[y(x,t;\ \mathbf{0})] - z(x,t)]\}\,dx\,dt$$
$$-\int_\Sigma v_1(\Sigma)\,z_1(\Sigma)\,d - \int_\Omega v_2(x)\,z_2(x)\,dx. \tag{3.16}$$

$$k = \int_Q \{R[y(x,t;\mathbf{0})] - z(x,t)\}^2\,dx\,dt + \int_\Sigma z_1(\Sigma)^2\,d + \int_\Omega z_2(x)^2\,dx. \tag{3.17}$$

PROOF. *Hypothesis on* $a(\mathbf{v},\mathbf{v})$.

(i) Evidently $a(\mathbf{u},\mathbf{v})$ is a bilinear form, since S is a linear evolution process. In addition,

$$y(x,t;\mathbf{v}) \in L^2(Q) \qquad \text{(Lemma 2.2)}.$$

Thus each term in (3.15) is bounded from above. Thus $a(\mathbf{v},\mathbf{v})$ is continuous in $\mathbf{v}$.

(ii) $a(\mathbf{v},\mathbf{v}) \geq \int_\Sigma v_1(\Sigma)^2\,d\Sigma + \int_\Sigma v_2{}^2(x)\,dx = \alpha\|\mathbf{v}\|^2_{L^2(\Sigma)\times L^2(\Omega)}, \qquad \alpha = 1.$

(iii) The symmetry is self-evident.

(iv) Again, linearity of $\mathrm{I}(\mathbf{v})$ is evident owing to the linearity of the evolution process S. Boundedness (and hence continuity) is obtained by application of the triangle and Schwartz inequalities:

$$|\mathrm{I}(\mathbf{v})| \leq |\int_Q R[y(x,t;\mathbf{v}) - y(x,t;\mathbf{0})]\{R[y(x,t;\mathbf{0})] - z(x,t)\}\,dx\,dt|$$
$$+ |\int_\Sigma v_1(\Sigma)\,z_1(\Sigma)\,d\Sigma| + |\int_\Omega v_2(x)\,z_2(x)\,dx|$$
$$\leq |M_0|\ \|R[y(x,t;\mathbf{v}) - y(x,t;\mathbf{0})]\|_{L^2(Q)} + |M|\ \|\mathbf{v}\|_{L^2(\Sigma)\times L^2(\Omega)}.$$

Since $y(x,t;\mathbf{v}) \in L^2(Q)$, $\mathbf{v} \in L^2(\Sigma)\times L^2(\Omega)$ (M_0 and M are positive constants), then $|\mathrm{I}(\mathbf{v})|$ is bounded and linear in $\mathbf{v}$, hence continuous in $\mathbf{v}$.

It is possible to consider many quadratic forms of the type (3.14). It is also possible to pose innocuous forms which resemble (3.14) superficially, but which violate the given hypothesis. Consider the following counter example due to Lions [16].

System S_c.

$$\frac{\partial y}{\partial t} + A[y] = f \quad \text{in } Q, \qquad Q = (0,1)\times(0,T].$$

$$y(0,t)=0;\ y(1,t)=\frac{1}{(T-t)^{\alpha}}, \qquad 2\alpha<1, \qquad u_1(\cdot)\in L^2(\Sigma).$$

$$y(x,0)=u_2(x), \qquad u_2(\cdot)\in L^2(\Omega).$$

$$A[y]=-\frac{\partial^2 y}{\partial x^2}(x,t)$$

Assertion. For the system S_c,

$$y(\cdot,T;\mathbf{u})\notin L^2(\Omega).$$

Proof. It has been shown (Lemma 2.2, Chapter 2) that $y(\cdot,\cdot)\in L^2(Q)$. Hence there exists a solution $y_m(x,t)$ defined by

$$y_m(x,t)=\sum_{i=1}^{m} y_i(t)w_i(x)$$

with the property that

$$\lim_{m\to\infty} y_m(x,t)\to y(x,t) \qquad \text{for almost every } x,\ t\in Q,$$

where $\{w_i(\cdot)\}_{i=1,2,3,\ldots}$ are a basis in $L^2(\Omega)$. It can be shown that:

$$y_i(t)=-\frac{dw_i(1)}{dx}\int_0^t e^{-\lambda_i(t-\tau)}u(1,t)\,d\tau, \tag{3.18}$$

$$|y_i(T)|=(\sqrt{2\Pi i})\int_0^T \frac{e^{-\lambda_i(T-\tau)}}{(T-\tau)^{\alpha}}d\tau. \tag{3.19}$$

Taking $v=\lambda_i(T-\tau)$, (3.19) becomes

$$|y_i(T)|=c\int_0^{\lambda_i T}\left[\frac{e^{-v}}{\lambda_i(1-\alpha)v^{\alpha}}\right]dv.$$

Noting the classical result that $\lambda_i\geq\lambda_1$, $i\geq 1$,

$$|y_i(T)|\geq\frac{c_i}{\lambda_i(1-\alpha)}\int_0^{\lambda_1 T}\frac{e^{-v}}{v^{\alpha}}dv=c_1\frac{i}{i^{2(1-\alpha)}},$$

$$\int_{\Omega} y_m(x,T)^2\,d\Omega=\sum_{i=1}^{m}|y_i(T)|^2\geq\frac{c_1{}^2}{i^{(2-4\alpha)}}$$

$$=\infty \qquad \text{if } 2-4\alpha\leq 1.$$

If an identification problem is posed for the system S_c with the following identification criteria:

$$J(\mathbf{v}) = \int_\Omega |y(x, T; v) - z(x, T)|^2 \, dx + \int_\Sigma [v_1(\Sigma) - z_1(\Sigma)]^2 \, d\Sigma + \int_\Omega [v_2(x) - z_2(x)]^2 \, dx, \tag{3.20}$$

then it is possible that no solution exists which minimizes (3.20) as was illustrated by the counter example. If we insists on constructing the error functional from measurements at the terminal time, $z(x, T)$, then $J(\mathbf{v})$ must be appropriately modified.

Lions has shown that for the system S_c,

$y(\cdot, T) \in H^{-1}(\Omega)$ (Dual space of the Sobolev space H_0^1 defined in Chapter 2, Section A.1).

Thus we choose

$$J(\mathbf{v}) = \|y(\cdot, T; \mathbf{u}) - z(\cdot, T)\|^2_{H^{-1}(\Omega)} + \|\mathbf{v}\|^2_{L^2(\Sigma) \times L^2(\Omega)}. \tag{3.21}$$

Where the norm in $H^{-1}(\Omega)$ is defined by the following operations:

for f, $g \in H^{-1}(\Omega)$,

solve

$$\left(-\frac{\partial^2}{\partial x^2} + 1\right)\phi = f, \qquad \phi = 0 \quad \text{on } \Gamma,$$

$$\left(-\frac{\partial^2}{\partial x^2} + 1\right)\Psi = g, \qquad \Psi = 0 \quad \text{on } \Gamma,$$

$$(f, g)_{H^{-1}(\Omega)} = \int_\Omega \left[\phi\Psi + \frac{\partial \Psi}{\partial x}\frac{\partial \phi}{\partial x}\right] dx.$$

This problem under the refinement implied by the minimization of (3.21) is well posed, and has a solution, considered in Chapter 4.

In summary, the state identification problem associated with the system S is that of minimizing the squared error arising from inaccurate measurements on S. This squared error is viewed as a quadratic functional of the boundary conditions $v_1(\Sigma)$ and initial data $v_2(x)$. The refinement of the incomplete and/or inexact measurements is accomplished by choosing v_1 and v_2 so that the quadratic functional is minimized along a trajectory of the system S. The conditions characterizing these refined estimates u_1 and u_2 are given in Chapter 4.

The term "refined estimate" as used here denotes a least squares fit of a trajectory of S to the given data I and O. Under certain hypotheses considered in Section A.3, this least squares refinement has a statistical significance.

A.3 STATISTICAL PROPERTIES OF THE REFINED ESTIMATE

In Section A.2.1 the measurement processes $z(\cdot,\cdot)$, $z_1(\cdot)$, and $z_2(\cdot)$ had associated with them the three error processes $\varepsilon(\cdot,\cdot)$, $\varepsilon_1(\cdot)$, and $\varepsilon_2(\cdot)$. No structure has been assumed for these processes in the definition of the refined estimate, except that they be $L^2(Q)$, $L^2(\Sigma)$, and $L^2(\Omega)$, respectively.

If statistical properties of these error processes are given, then the data refinement problem can be phrased statistically, *but need not be*.

A special case is of particular interest. Consider the system S with the additional statistical data:

Hypothesis on $\varepsilon(x,t)$, $\varepsilon_1(\Sigma)$, *and* $\varepsilon_2(x)$: Given test tunctions $\phi(x)$ and $\phi_1(\Gamma)$, define:

$$\int_\Omega \phi(x)\varepsilon(x,t)\,dx = n(t), \qquad \phi(\cdot)\in L^2(\Omega),$$

$$\int_\Gamma \phi_1(\Gamma)\varepsilon_1(\Sigma)\,d\Gamma = n_1(t), \qquad \phi_1(\cdot)\in {}^2L(\Gamma),$$

$$\int_\Omega \phi(x)\varepsilon_2(x)\,d\Omega = n_2(0), \qquad \phi_2(\cdot)\in L^2(\Gamma).$$

$$\begin{aligned} E[n(t)] &= 0 \qquad \text{for all } t\in(0,T], \\ E[n_1(t)] &= 0 \qquad \text{for all } t\in(0,T], \\ E[n_2(0)] &= \bar{n}_2 \end{aligned}$$

$$\begin{aligned} E[n(t)n(\tau)'] &= V(t,\tau) = I(t-\tau), \\ E(n_1(t)n_1(\tau)'] &= R(t,\tau) = I(t-\tau), \\ E[n_2 n_2'] &= V_0 \quad = I. \end{aligned}$$

The correlation coefficient for any of the pairs (n,n_1), (n_1,n_2), (n,n_2) is zero. Let n, n_1, and n_2 be Gaussianly distributed.

It is important to notice that the error processes $\varepsilon(x, t)$ and $\varepsilon_1(\Sigma)$ at some time $t \in (0, T]$ *are not* random functions of their spatial arguments. Another important property of the structure imposed is that integrals such as

$$\int_0^T f(t) n(t)\, dt$$

must be taken with respect to some probablistic measure. No attempt at rigor is made here. We observe these technicalities (which are of considerable importance, but not in this context) and proceed formally.

In this new statistical framework the state identification problem is phrased as follows:

$$\underset{y^*(\cdot,\cdot)}{\text{maximize}}\ P[y^* - \Delta(\cdot,\cdot) \le y^* \le y^* + \Delta(\cdot\,.\cdot) \,|\, I, O],$$

where $P[y|I, O]$ denotes the probability that the event y occurs given that I and O have.

The character of the hypothesis and the linearity ot S cause this conditional probability to have a Gaussian density. Owing to the nonnegativity of probability density functions, maximizing the probability is the same as maximizing the probability density. Thus we have that

$$\max_{y^*} P[y^* - \Delta \le y^* \le y^* + \Delta \,|\, I, O] = \int_{y^*-\Delta}^{y^*+\Delta} \max p(y|I, O)\, dy .$$

For small Δ, $\Rightarrow \max\limits_{y^*} p(y^*|I, O)$.

Recall that $p(y^*|I,O)$ is Gaussian. Then the indicated *maximization* is accomplished by *minimizing* the argument of the exponential expression:

$$\max_{y^*} p(y^*|I, O) \Rightarrow \max_{y^* \in S} k_y \exp(-\tfrac{1}{2}[J(y^*)]) \tag{3.22}$$

$$\Rightarrow \min_{y^* \in S} J(y^*) .$$

In (3.22), $J(y^*)$ is a quadratic functional, and thus minimizing (2.22) at each instant t is equivalent to minimizing the integral

$$\min_{y^* \in S} \int_0^T J(y^*)\, dt , \tag{3.23}$$

where

$$J(y^*) = (y^*(\cdot,\cdot) - z(\cdot,\cdot), V^{-1}[y^*(\cdot,\cdot) - z(\cdot,\cdot)])_{L^2(\Omega)}$$
$$+ (u_1{}^* - z_1(\cdot,\cdot), R[u_1{}^* - z_1(\cdot,\cdot)])_{L^2(\Gamma)}$$
$$+ (u_2{}^* - z_2(\cdot), V_0^{-1}[u_2{}^* - z_2(\cdot)])_{L^2(\Omega)} . \tag{3.24}$$

REMARKS. (i) The integral in (3.23) is taken in the stochastic sense.

(ii) $J(y^*)$ given by (3.24) was obtained by approximating $y^*(x, t)$ in $L^2(Q)$ and then applying the results of Pearson [19]. We returned to the infinite dimensional case formally. Note that this approach is valid since $y(x, t) \in L^2(Q)$. See Appendix 3.1.

(iii) The statistical problem is evidently equivalent to the deterministic problem of Section A.2 under the specialized hypothesis given. Thus the data refinement problem with least squares criteria is not without statistical respectability.

In the sequel, further statistical connotation or framework is avoided in favor of a purely variational approach to the characterization of extremals to the quadratic functional of (3.14) lying in S. We now turn our attention to such characterizations, carried out in Chapter 4.

REMARKS. The deterministic problem referred to is the one in which $z(x, t)$ is given by (3.10). The case in which measurements are taken at discrete points in the spatial domain has not been investigated. The case of no measurements in the interior region Q corresponds to an infinite variance matrix V.

CHAPTER 4

CHARACTERIZATION OF THE SOLUTION OF THE STATE INDENTIFICATION PROBLEM

A.0 INTRODUCTION

The state identification problem has been posed as a variational one, that of minimizing a quadratic functional along given system state trajectories, each parameterized by the initial and boundary data, denoted by **v**. The indicated minimization is accomplished by the appropriate choice of an "admissible" **v**. The purpose of this chapter is to indicate mathematically the "appropriate" choice of **v** (denoted by **u**), for a variety of systems, associated measurement processes, and quadratic error functionals. In particular, **u** will be characterized as the solution to a set of simultaneous equations.

Theorems establishing the existence and uniqueness of solutions to the various identification problems are given in each case.

The task of actually constructing a solution **u** from the characterization given by the simultaneous equations referred to in this introduction is the domain of Chapter 5.

A.1 PARABOLIC SYSTEMS, $V = L^2(\Sigma) \times L^2(\Omega)$

This section contains the solution of the state identification problem for parabolic systems where the output measurements are:

(A) $z(x,t) = y(x,t)+\varepsilon(x,t), \qquad z(\cdot,\cdot) \in L^2(Q)\,,$

(B) $z(x^i,t) = y(x^i,t)+\varepsilon(x^i,t), \qquad z(x^i,\cdot) \in L^2(0,T)\,,$

(C) $z(x,t) = 0\,.$

Additionally it is assumed that the admissible set of boundary and initial data, denoted by V, is the entire space $L^2(\Sigma)\times L^2(\Omega)$. In each problem induced by (A), (B), and (C), the measurement on the boundary and at the time are (respectively):

$$z_1(\Sigma) = u_1(\Sigma)+\varepsilon_1(\Sigma), \qquad z_1(\Sigma) \in L^2(\Sigma)\,,$$
$$z_2(x) = u_2(x)+\varepsilon_2(x), \qquad z_2(x) \in L^2(\Omega)\,.$$

We now consider (in turn) the identification problem induced by (A), (B), and (C) for each of the following parabolic systems:

(i) Dirichlet boundary conditions,
(ii) Neumann boundary conditions,
(iii) "mixed" boundary conditions.

Remarks. It is assumed that $z(x,t)$ is measured everywhere in Q, a physically unrealistic situation.

$z(x^i,t)$ corresponds to pointwise measurements located at the spatial coordinates x^i, $i=1,2,\ldots,\nu$.

$z(x,t)=0$ implies that no interior measurements are taken.

In Section A.3 we consider the case where V is a closed convex subset of $L^2(\Sigma)\times L^2(\Omega)$.

PR.I: *Dirichlet Boundary Conditions*

System S_D.

$$\frac{\partial y(x,t)}{\partial t}+A[y(x,t)]=f(x,t)\quad \text{in } Q, \tag{4.1}$$

$$y(\Sigma)=v_1(\Sigma)\qquad \text{on } \Sigma, \tag{4.2}$$

$$y(x,0)=v_2(x)\qquad \text{in } \Omega. \tag{4.3}$$

Hypothesis on the System, H_D.

$$f(x,t)\in L^2(Q):\qquad \int_0^T\int_\Omega |f(x,t)|^2\,dx\,dt<\infty,$$

$$\left.\begin{array}{ll} v_1(\Sigma)\in L^2(\Sigma): & \displaystyle\int_0^T\int_\Gamma v_1(s,t)|\,^2 ds\,dt <\infty \\ v_2(x)\in L^2(\Omega): & \displaystyle\int_\Omega |v_2(x)|^2\,dx <\infty \end{array}\right\}=V.$$

Problem PR.IA. Let

$$J(\mathbf{u})=\inf_{\mathbf{v}\in V} J(\mathbf{v}),$$

where

$$J(\mathbf{v}) = \left\{ \int_Q [y(x,t;\mathbf{v}) - z(x,t)]^2 \, dx \, dt \right.$$

$$+ \int_\Sigma [v_1(\Sigma) - z_1(\Sigma)]^2 \, d\Sigma$$

$$+ \int_\Omega [v_2(x) - z_2(x)]^2 \, dx \} . \tag{4.4}$$

It is required to find a characterization of $\mathbf{u}$ which has the property that $J(\mathbf{u}) = \mathrm{Inf}_{\mathbf{v} \in V} J(\mathbf{v})$ for any $\mathbf{v} \in V$ along a trajectory $y(\cdot, \cdot, \mathbf{u})$ of the system S_D under the hypothesis H_D.

Problem PR.IB. Let

$$J(\mathbf{u}) = \inf_{\mathbf{v} \in V} J(\mathbf{v}),$$

$$J(\mathbf{v}) = \sum_{i=1}^{\nu} \left\{ \int_0^T [|y(x^i, t; \mathbf{v}) - z(x^i, t)|^2] \, dt \right.$$

$$+ \int_\Sigma [v_1(\Sigma) - z_1(\Sigma)]^2 \, d\Sigma$$

$$\left. + \int_\Omega [v_2(x) - z_2(x)]^2 \, dx \right\}. \tag{4.5}$$

It is required to find a characterization of $\mathbf{u}$ which has the property $J(\mathbf{u}) = \mathrm{Inf}_{\mathbf{v} \in V} J(\mathbf{v})$ for any $v \in V$ along trajectories $y(x^i, ., \mathbf{u})$ of the system S_D under the hypothesis H_D.

Problem PR.IC.

$$J(\mathbf{v}) = \int_\Sigma [v_1(\Sigma) - z_1(\Sigma)]^2 \, d\Sigma$$

$$+ \int_\Omega [v_2(x) - z_2(x)]^2 \, dx . \tag{4.6}$$

The solution to each of the three problems PR.AI, PR.IB, and PR.IC stated in conjunction with the state evolution process S_D is given by the Theorem 4.1,IA, Theorem 4.1,IB, and Theorem 4.1,IC which follow:

THEOREM 4.1,IA. *Given the system S_D, hypothesis H_D,* and the functional $J(\mathbf{v})$ of PR.IA, then:

(i) *There exists one and only one $u \in V$ with the property that*

$$J(\mathbf{u}) \leq J(\mathbf{v}) \quad \text{for all } \mathbf{v} \in V .$$

(ii) *The boundary and initial data* $\mathbf{u}$ *are uniquely characterized* by the simultaneous solution of the following system of equations:

$$\frac{\partial y(x,t;\mathbf{u})}{\partial t} + A[y(x,t;\,\mathbf{u})] = f(x,t) \quad \text{in } Q, \tag{4.7}$$

$$y(\Sigma) = u_1(\Sigma) \quad \text{on } \Sigma, \tag{4.8}$$

$$y(x,0) = u_2(x) \quad \text{in } \Omega. \tag{4.9}$$

$$-\frac{\partial p(x,t;\mathbf{u})}{\partial t} + A[p(x,t;\mathbf{u})] = y(x,t;\,\mathbf{u}) - z(x,t) \quad \text{in } Q, \tag{4.10}$$

$$p(\Sigma) = 0 \quad \text{on } \Sigma, \tag{4.11}$$

$$p(x,T) = 0 \quad \text{in } \Omega. \tag{4.12}$$

$$-\frac{\partial p(\Sigma)}{\partial \nu_{A^*}} + u_1(\Sigma) - z_1(\Sigma) = 0 \quad \text{on } \Sigma, \tag{4.13}$$

$$p(x,0) + u_2(x) - z_2(x) = 0 \quad \text{in } \Omega. \tag{4.14}$$

THEOREM 4.1,IB. *Given the system S_D, hypothesis H_D, and the functional* $J(\mathbf{v})$ *of* PR.IB, *then*:

(i) *There exists one and only one* $\mathbf{u} \in V$ *with the property that*

$$J(\mathbf{u}) \leq J(\mathbf{v}) \qquad \text{for all } v \in V .$$

(ii) *The boundary and initial data* $\mathbf{u}$ *are uniquely characterized by the simultaneous solution of the following system of equations*:

$$\frac{\partial y(x,t;\mathbf{u})}{\partial t} + A[y(x,t;\mathbf{u})] = f(x,t) \quad \text{in } Q, \tag{4.15}$$

$$y(\Sigma) = u_1(\Sigma) \quad \text{on } \Sigma, \tag{4.16}$$

$$y(x,0 = u_2(x) \quad \text{in } \Omega. \tag{4.17}$$

$$-\frac{\partial p(x,t;\mathbf{u})}{\partial t}+A[p(x,t;\mathbf{u})=\sum_{i=1}^{v}[y(x^i,t;\mathbf{u})-z(x^i,t)]\delta(x-x^i), \quad (4.18)$$

$$p(\Sigma)=0, \quad (4.19)$$

$$p(x,T)=0 \quad (4.20)$$

($\delta(x-x^i)$ *is the Dirac delta function*).

$$-\frac{\partial p(\Sigma)}{\partial v}+u_1(\Sigma)-z_1(\Sigma)=0, \quad (4.21)$$

$$p(x,0)+u_2(x)-z_2(x)=0. \quad (4.22)$$

THEOREM 4.1,IC. *Given the system* S_D, *hypothesis* H_D, *and the functional* $J(\mathbf{v})$ *of* PR.IC; *then*:

(i) *There exists one and only one* $\mathbf{u}\in V$ *with the property that*

$$J(\mathbf{u})\le J(\mathbf{v}) \qquad \text{for all } v\in V.$$

(ii) *The unique solution for* $\mathbf{u}$ *is trivial and is given by*:

$$u_1(\Sigma)=z_1(\Sigma), \quad (4.23)$$

$$u_2(x)=z_2(x). \quad (4.24)$$

These three theorems are now proved in turn. Before proceeding, however, the following lemmas which assist in the proofs are presented:

LEMMA 4.1. *There exists a unique solution* $y(x,t;\mathbf{v})$ *to the system* S_D *and hypothesis* H_D *with the property that*

$$y(\cdot,\cdot;\mathbf{v})\in L^2(Q).$$

This lemma was stated earlier, in Chapter 2, namely, Lemma 2.2.

LEMMA 4.2. *For* $y(x,t;\mathbf{v}), z(x,t;\mathbf{v})\in L^2(Q)$, *then there exists one and only one solution to the system of equations* (4.10), (4.11), *and* (4.12) *with the properties*

(i) $p(\cdot,\cdot;\mathbf{v})\in L^2(Q)$,

(ii) $\dfrac{\partial p(\cdot)}{\partial v_{A^*}}\in L^2(\Sigma)$.

REMARK: In fact, a stronger result is possible (i.e., it is possible to restrict $p(x,t;\mathbf{v})$ to the Sobolev space H_0^1, H_0^2, etc.). See [20].

PROOF OF LEMMA 4.2. *Part* (i) *of the lemma is the same as Lemma* 4.1, *simply reversing time from* $t = T$ *to* $t = 0$. *Part* (ii) *is due to Lions and Magenes* [20].

LEMMA 4.3. *For* $y(x^i, \cdot; \mathbf{v})$ *and* $z(x^i, \cdot; \mathbf{v}) \in L^2(0, T)$, *then there exists one and only one solution to the system of equations* (4.18), (4.19), *and* (4.20) *with the properties*

(i) $p(\cdot, \cdot; \mathbf{v}) \in L^2(Q)$,

(ii) $\dfrac{\partial p(\cdot)}{\partial v_{A^*}} \in L^2(\Sigma)$.

PROOF OF LEMMA 4.3. Same as for Lemma 4.2.

LEMMA 4.4. *The functional* $J(\mathbf{v})$ *of* PR.IA *has the representation*

$$J(\mathbf{v}) = a(\mathbf{v}, \mathbf{v}) - 2\,\mathrm{I}(\mathbf{v}) + c, \tag{4.25}$$

where

(i) $a(\mathbf{w}, \mathbf{v})$ *is a continuous bilinear form on* $L^2(\Sigma) \times L^2(\Omega)$,
(ii) $\mathrm{I}(\mathbf{v})$ *is a continuous linear form on* $L^2(\Sigma) \times L^2(\Omega)$,
(iii) c *is a nonnegative constant.*

PROOF OF LEMMA 4.4.If we define

$$a(\mathbf{v}, \mathbf{v}) = \int_Q [y(x, t; \mathbf{v}) - y(x, t; \mathbf{0})]^2\, dx\, dt + \int_\Sigma [v_1(\Sigma)]^2\, d\Sigma + \int_\Omega [v_2(x)]^2\, dx, \tag{4.26}$$

$$\mathrm{I}(\mathbf{v}) = -\left\{\int_Q [y(x, t; \mathbf{v}) - y(x, t; \mathbf{0})][y(x, t; \mathbf{0}) - z(x, t)]\, dx\, dt - \int_\Omega v_1(\Sigma) z_1(\Sigma)\, d\Sigma - \int_\Omega v_2(x) z_2(x)\, dx\right\}, \tag{4.27}$$

$$c = \int_Q [y(x, t; \mathbf{0}) - z(x, t)]^2\, dx\, dt + \int_\Sigma [z_1(\Sigma)]^2\, d\Sigma + \int_\Omega [z_2(x)]^2\, dx. \tag{4.28}$$

Then is is clear that:

$$J(\mathbf{v}) = a(\mathbf{v}, \mathbf{v}) - 2\mathrm{I}(v) + c .$$

It remains to be shown that $a(\mathbf{v}, \mathbf{v})$ given by (4.26) and $\mathrm{I}(\mathbf{v})$ given by (4.27) are continuous bilinear and linear forms (respectively) on $L^2(\Sigma) \times L^2(\Omega)$.

Owing to the linearity of the process S_D, bilinearity of $a(\mathbf{v}, \mathbf{v})$ and linearity of $\mathrm{I}(\mathbf{v})$ are evident. Continuity of $a(\mathbf{v}, \mathbf{v})$ and $\mathrm{I}(\mathbf{v})$ is immediate from the boundedness of (4.26) and (4.27) arising from the hypothesis $H4$ and Lemma 4.1. Finally, it is clear from (4.26) that $a(\mathbf{v}, \mathbf{v})$ is coercive, that is,

$$a(\mathbf{v}, \mathbf{v}) \geq \alpha \|\mathbf{v}\|^2_{L^2(\Sigma) \times L^2(\Omega)} \qquad \text{for } \alpha = 1 .$$

LEMMA 4.5. *The functional $J(\mathbf{v})$ of* PR.IB *has the representation*

$$J(\mathbf{v}) = a(\mathbf{v}, \mathbf{v}) - 2\mathrm{I}(\mathbf{v}) + c , \tag{4.29}$$

where

(i) *$a(\mathbf{v}, \mathbf{w})$ is a continuous bilinear form on $L^2(\Sigma) \times L^2(\Omega)$,*
(ii) *$\mathrm{I}(\mathbf{v})$ is a continuous linear form on $L^2(\Sigma) \times L^2(\Omega)$,*
(iii) *c is a real nonnegative constant.*

PROOF OF LEMMA 4.5. Same as for Lemma 4.4.

LEMMA 4.6. *The functional $J(\mathbf{v})$ of* PR.IC *has the representation*

$$J(\mathbf{v}) = a(\mathbf{v}, \mathbf{v}) - 2\mathrm{I}(\mathbf{v}) + c ,$$

where

(i) *$a(\mathbf{v}, \mathbf{w})$ is a continuous bilinear form on $L^2(\Sigma) \times L^2(\Omega)$,*
(ii) *$\mathrm{I}(\mathbf{v})$ is a continuous linear form on $L^2(\Omega) \times L^2(\Omega)$,*
(iii) *c is a real nonnegative constant.*

PROOF OF LEMMA 4.6. The proof is the same as for Lemma 4.4.

The groundwork for the proof of Theorems 4.1,IA, 4.1,IB, 4.1,IC has been laid. We now proceed directly to the proofs.

PROOF OF THEOREM 4.1,IA. Invoke Lemma 4.4 and write

$$J(\mathbf{v}) = a(\mathbf{v}, \mathbf{v}) - 2\mathrm{I}(\mathbf{v}) + c .$$

with $V = L^2(\Sigma) \times L^2(\Omega)$, we have that

(a) V is convex,
(b) V is closed.

Then, all the hypotheses of Theorem 2.5 are satisfied and thus:

(i) There exists one and only one $\mathbf{u} \in V$ such that

$$J(\mathbf{u}) \leq J(\mathbf{v}) \qquad \text{for all } \mathbf{v} \in V ,$$

Furthermore, **u** is characterized by

$$a(\mathbf{u},\mathbf{v}) - \mathrm{I}(\mathbf{v}) = 0 \qquad \text{for all } \mathbf{v} \in V. \tag{4.30}$$

Using the definitions (4.26) and (4.27), (4.30) is

$$\int_Q [y(x,t;\mathbf{u}) - z(x,t)][y(x,t;\mathbf{v}) - y(x,t;\mathbf{0})]\,dx\,dt$$

$$+ \int_\Sigma [u_1(\Sigma) - z_1(\Sigma)]v_1(\Sigma)\,d\Sigma + \int_\Omega [u_2(x) - z_2(x)]v_2(x)\,dx = 0. \tag{4.31}$$

The system adjoint to S_D evolves according to

$$-\frac{\partial p(x,t;\mathbf{u})}{\partial t} + A[p(x,t;\mathbf{u})] = y(x,t;\mathbf{u}) - z(x,t), \tag{4.32}$$

$$p(\Sigma) = 0, \tag{4.33}$$

$$p(x,T) = 0. \tag{4.34}$$

If (4.32) is multiplied through by $[y(x,t;\mathbf{v}) - y(x,t;\mathbf{0})]$, $\mathbf{v} \in V$ (arbitrary) and the result integrated over Q, we obtain:

$$\int_Q \left\{\left[-\frac{\partial p(x,t;\mathbf{u})}{\partial t} + A[p(x,t;\mathbf{u})]\right][y(x,t;\mathbf{v}) - y(x,t;\mathbf{0})]\right\}dx\,dt$$

$$= \int_Q [y(x,t;\mathbf{u}) - z(x,t)][y(x,t;\mathbf{v}) - y(x,t;\mathbf{0})]\,dx\,dt. \tag{4.35}$$

The reason for introducing the adjoint system (4.32) to (4.34) and then generating (4.35) is now apparent: The R. H. S. of (4.35) is the same as the first term in (4.31). Thus the characterizarion of **u** given by (4.30) is equivalent, via (4.31) and (4.35), to:

$$\int_Q \left\{\left[-\frac{\partial p(x,t;\mathbf{u})}{\partial t} + A[p(x,t;\mathbf{u})]\right][y(x,t;\mathbf{v}) - y(x,t;\mathbf{0})]\right\}dx\,dt$$

$$+ \int_\Sigma [u_1(\Sigma) - z_1(\Sigma)]v_1(\Sigma)\,d\Sigma + \int_\Omega [u_2(x) - z_2(x)]v_2(x)\,dx = 0$$

$$\text{for all } \mathbf{v} \in V. \tag{4.36}$$

The result which is desired (equations (4.13) and (4.14)) is at hand.

Using Green's theorem (see Remark 4.1) for "integration by parts," observe that

$$\int_Q A[p(x,t;\mathbf{u})]\,[y(x,t;\mathbf{v})-y(x,t;\mathbf{0})]\,dx\,dt$$

$$= \int_Q p(x,t;\mathbf{u})\left[\frac{\partial y(x,t;\mathbf{0})}{\partial t} - \frac{\partial y(x,t;\mathbf{v})}{\partial t}\right] dx\,dt$$

$$+ \int_\Sigma \left\{ -\frac{\partial p(\Sigma;\mathbf{u})}{\partial \nu_{A^*}}\,\mathbf{v}_1(\Sigma) + p(\Sigma;\mathbf{u})\left[\frac{\partial y(\Sigma;\mathbf{v})}{\partial \nu_A} - \frac{\partial y(\Sigma;\mathbf{0})}{\partial \nu}\right]\right\} d\Sigma. \tag{4.37}$$

In addition,

$$\int_Q -\frac{\partial p(x,t;\mathbf{u})}{\partial t}\,[y(x,t;\mathbf{v})-y(x,t;\mathbf{0})]\,dx\,dt$$

$$= \int_Q -p(x,t;\mathbf{u})\left[\frac{\partial y(x,t;\mathbf{0})}{\partial t} - \frac{\partial y(x,t;\mathbf{v})}{\partial t}\right] dx\,dt$$

$$+ \int_\Omega \{p(x,0;\mathbf{u})\,[v_2(x)] - p(x,T;\mathbf{u})\,[y(x,T;\mathbf{v})]\}\,dx. \tag{4.38}$$

Putting (4.37) and (4.38) into (4.36) where appropriate and noting that $p(x,T;\mathbf{u}) = p(\Sigma,\mathbf{u}) = 0$, we obtain:

$$\int_\Sigma \left\{\left[-\frac{\partial p(\Sigma;\mathbf{u})}{\partial \nu_{A^*}} + u_1(\Sigma) - z_1(\Sigma)\right] v_1(\Sigma)\,d\Sigma\right.$$

$$+ \int_\Omega \{[p(x,0;\mathbf{u}) + u_2(x) - z_2(x)]\,v_2(x)\,dx = 0 \qquad \text{for all } \mathbf{v}\in V. \tag{4.39}$$

Finally, the arbitrariness of $v_1(\Sigma)$ and $v_2(x)$ give the required result:

$$-\frac{\partial p(\Sigma;\mathbf{u})}{\partial \nu_{A^*}} + u_1(\Sigma) - z_1(\Sigma) = 0, \tag{4.40}$$

$$p(x,0;\mathbf{u}) + u_2(x) - z_2(x) = 0. \tag{4.41}$$

Assertion (ii) of Theorem 4.1,IA has been established, since (by Lemmas 4. and 4.2) the solution of the system of equations (4.7) through (4.14) is unique.

Remark 4.1. Consider the application of Green's theorem in the following situation:

$$\int_Q A[p(x,t;\mathbf{u})][y(x,t;\mathbf{v})-y(x,t;\mathbf{0})]\,dx\,dt$$

$$= \int_Q A[y(x,t;\mathbf{v})-y(x,t;0)]\,p(x,t;\mathbf{u})\,dx\,dt$$

$$+ \int_\Sigma \left\{ -\frac{\partial p(\Sigma;\mathbf{u})}{\partial v_{A^*}} v_1(\Sigma) - p(\Sigma;\mathbf{u}) \left[\frac{\Sigma y(\Sigma;\mathbf{v})}{\partial v} - \frac{y(\Sigma;\mathbf{0})}{\partial v} \right] \right\} d\Sigma . \qquad (4.42)$$

It is assumed that $p(x,t;\mathbf{u})$ is sufficiently differentiable $(\partial p/\partial x^i) \in L^2(Q)$, $(\partial^2 p/\partial x^{i2} \in L^2(Q))$ so that the differentiation of $[y(x,t;\mathbf{v})-y(x,t;0)]$ can be taken in the distribution sense. This remark is essential, since $A[y(x,t;\mathbf{v})] \notin L^2(Q)$.

The proof of Theorem 4.1,IA has been given in complete detail. The proofs of Theorem 4.1,IB and 4.1,IC proceed in precisely the same way. Thus we give a formal outline of the salient features of the proof in each case.

Proof of Theorem 4.1, IB. The proof of Part (i) is the same as given in the proof of Theorem 4.1,IA. The structure of the proof of Part (ii) is (again) the same as given in the proof of Theorem 4.1,IA. As before, the existence of an appropriately smooth $p(x,t;\mathbf{u})$ sarisfying (4.17), (4.18), and (4.19) is again (as in Remark 4.1) assumed. The assumption is vindicated by the results of Lions [20].

Proof of Theorem 4.1,IC. Part (i) carries over exactly as before. Part (ii) is trivial, for we have

$$-\frac{\partial p(x,t;\mathbf{u})}{\partial t} + A[p(x,t;\mathbf{u})] = 0,$$

$$p(\Sigma;\mathbf{u}) = 0,$$

$$p(x,T) = 0,$$

which implies that

$$p(x,t;\mathbf{u}) \equiv 0 \quad \text{everywhere},$$

which, in turn, implies

$$u_1(\Sigma) = z_1(\Sigma) \quad \text{almost everywhere in } L^2(\Sigma),$$

$$u_2(x) = z_2(x) \quad \text{almost everywhere in } L^2(\Omega).$$

The solution of PR.I is complete. Pr.II and PR.III differ from PR.I in that the boundary conditions of the parabolic system S are Neumann and "mixed," respectively. The criteria of identification given for PR.II and PR.III are the same as for PR.I. We now define PR.II, the identification problem for the parabolic system with Neumann boundary conditions:

PR.II. *Neumann Boundary Conditions*

$$\frac{\partial y(x,t)}{\partial t} + A[y(x,t)] = f(x,t) \quad \text{in } Q, \tag{4.43}$$

$$\frac{\partial y(\Sigma)}{\partial \nu} = v_1(\Sigma) \quad \text{on } \Sigma, \tag{4.44}$$

$$y(x,0) = v_2(x) \quad \text{in } \Omega. \tag{4.45}$$

Hypothesis on the System, H_N.

$$f(.,\cdot) \in L^2(Q): \quad \int_Q |f(x,t)|^2 \, dx \, dt < \infty,$$

$$\left.\begin{aligned} v_1(\cdot) \in L^2(\Sigma): &\quad \int_\Sigma |v_1(\Sigma)|^2 \, d\Sigma < \infty \\ v_2(\cdot) \in L^2(\Omega): &\quad \int_\Omega |v_2(x)|^2 \, dx < \infty \end{aligned}\right\} = V.$$

For the system S_N and hypothesis H_N, there are the three identification problems, PR.II.A, PR.I.B, and PR.I.C., modified in each case by the statement:

"where, for given $\mathbf{v} \in V$, $y(x,t;\mathbf{v})$ is *the* trajectory of the system S_N under the hypothesis H_N."

The characterization of $\mathbf{u}$, the solution to each problem, is given in the following three theorems:

THEOREM 4.1,IIA. *Given the system* S_N, *hypothesis* H_N, *and the functional* $J(\mathbf{v})$ *of* PR.IIA, *then*:

(i) *There exists one and only one* $\mathbf{v} \in V$ *with the property that*

$$J(\mathbf{u}) \leq J(\mathbf{v}) \quad \text{for all } \mathbf{v} \in V.$$

(ii) *The boundary and initial data* **u** *is uniquely characterized by the simultaneous solution of the following system of equations*:

$$\frac{\partial y(x,t;\mathbf{u})}{\partial t} + A[k(x,t;\mathbf{u})] = f(x,t) \quad \text{in } Q, \tag{4.46}$$

$$\frac{\partial y(\Sigma)}{\partial \nu} = u_1(\Sigma) \quad \text{on } \Sigma, \tag{4.47}$$

$$y(x,0) = u_2(x) \quad \text{in } \Omega. \tag{4.48}$$

$$-\frac{\partial p(x,t;\mathbf{u})}{\partial t} + A[p(x,t;\mathbf{u})] = y(x,t;\mathbf{u}) - z(x,t) \quad \text{in } Q, \tag{4.49}$$

$$\frac{\partial p(\Sigma)}{\partial \nu_{A^*}} = 0 \quad \text{on } \Sigma, \tag{4.50}$$

$$p(x,T) = 0 \quad \text{in } \Omega, \tag{5.51}$$

$$p(\Sigma;\mathbf{u}) + u_1(\Sigma) - z_1(\Sigma) = 0 \quad \text{on } \Sigma, \tag{4.52}$$

$$p(x,0;\mathbf{u}) + u_2(x) - z_2(x) = 0 \quad \text{in } \Omega. \tag{5.53}$$

PROOF OF THEOREM 4.1,IIA. The proof of part (i) is the same for part (i) of Theorem 4.1,IA, modified in the following way: Lemma 4.1 applies to S_N and H_N.

The proof of part (ii) is the same as the proof of part (ii) of Theorem 4.1,IA.

THEOREM 4.1,IIB. *Given the system* S_N, *hypothesis and the* H_N, *functional* $J(\mathbf{v})$ of PR.IIB, *then*:

(i) *There exists one and only one* $\mathbf{u} \in V$ *with the property that*

$$J(\mathbf{u}) \leq J(\mathbf{v}) \quad \text{for all } \mathbf{v} \in V.$$

(ii) *The boundary and initial data* **u** *is uniquely characterized by the simultaneious solution of the following system of equations*:

$$\frac{\partial y(x,t;\mathbf{u})}{\partial t} + A[y(x,t;\mathbf{u})] = f(x,t) \quad \text{in } Q, \tag{4.54}$$

$$\frac{\partial y(\Sigma)}{\partial \nu} = u_1(\Sigma) \quad \text{on } \Sigma, \tag{4.55}$$

$$y(x,0) = u_2(x) \quad \text{in } \Omega. \tag{4.56}$$

$$-\frac{\partial p(x,t;\mathbf{u})}{\partial t}+A[p(x,t;\mathbf{u})]=\sum_{i=1}^{\nu}[y(x^i,t;\mathbf{u})-z(x^i,t)]\delta(x-x^i), \quad (4.57)$$

$$\frac{\partial p(\Sigma;\mathbf{u})}{\partial \nu_{A^*}}=0, \quad (4.58)$$

$$p(x,T)=0, \quad (4.59)$$

$$p(\Sigma;\mathbf{u})+u_1(\Sigma)-z_1(\Sigma)=0, \quad (4.60)$$

$$p(x,0;\mathbf{u})+u_2(x)-z_2(x)=0. \quad (4.61)$$

PROOF OF THEOREM 4.1,IIB. The proof of part (i) is the same as for part (i) of Theorem 4.1,IB, modified in the following way: Lemma 4.1 applies to S_N and H_N. The proof of part (ii) is the same as for part (ii) of Theorem 4.1,IB.

THEOREM 4.1,IIC. *Given the system S_N, hypothesis H_N, and the functional $J(\mathbf{v})$ of* PR.IIC, *then*:

(i) *There exists one and only one $u \in V$ with the property that*

$$J(\mathrm{u}) \leq J(\mathbf{v}) \quad \text{for all } \mathbf{v} \in V.$$

(ii) *The unique solution for* **u** *is trivial and is given by*

$$u_1(\Sigma)=z_1(\Sigma), \quad (4.62)$$

$$u_2(x)=z_2(x). \quad (4.63)$$

PROOF OF THEOREM 4.1,IIC. The proof of part (i) is the same as for the proof of part(i) of Theorem 4.1,IC, with the modification: Lemma 4.1 applies to S_N and H_N. Part (ii) is the same as for part (ii) of Theorem 4.1,IC.

The last parabolic system type to be considered is the parabolic system with "mixed" boundary conditions. The associated identification problems PR.IIIA to PR.IIIC lead to results for the unique choice of **u**, the optimal value of the boundary and initial data for the given problem. As before, the characterization is given as a theorem. In this case, however, the proofs will be omitted as they are obvious in the light of preceding work.

PR.III. *"Mixed" Boundary Conditions*

System S_M.

$$\frac{\partial y(x,t)}{\partial t}+A[y(x,t)]=f(x,t) \quad \text{in } Q, \quad (4.64)$$

$$\frac{\partial y(\Sigma)}{\partial \nu_A}+\alpha(\Sigma)y(\Sigma)=v_1(\Sigma) \quad \text{on } \Sigma, \quad (4.65)$$

$$y(x,0)=v_2(x) \quad \text{in } \Omega. \quad (4.66)$$

Hypothesis H_M.

$$f(\cdot,\cdot)\in L^2(Q):\quad \int_Q |f(x,t)|^2\,dx\,dt < \infty$$

$$\left.\begin{array}{ll} u_1(.)\in L^2(\Sigma): & \displaystyle\int_\Sigma |v_1(\Sigma)|^2\,d\Sigma \quad < \infty \\ u_2(\cdot)\in L^2(\Omega): & \displaystyle\int_\Omega |v_2(x)|^2\,dx \quad < \infty \end{array}\right\} = V .$$

In addition,

$$\alpha(\Sigma) \geq 0 .$$

The theorems characterizing the optimal $\mathbf{v}\in V$ now follow:

THEOREM 4.1,IIIA. *Given the system* S_M *and hypothesis* H_M *and the functional* $J(\mathbf{v})$ *of* PR.IIIA; *then*:

(i) *There exists one and only one* $\mathbf{u}\in V$ *with the property that*

$$J(\mathbf{u}) \leq J(\mathbf{v}) \quad \text{for all } \mathbf{v}\in V .$$

(ii) *The boundary and initial data* **u** *is uniquely characterized by the simultaneous solution of the following system of equations*:

$$\frac{\partial y(x,t;\mathbf{u})}{\partial t} + A[y(x,t;\mathbf{u})] = f(x,t) \quad \text{in } Q, \tag{4.67}$$

$$\frac{\partial y(\Sigma)}{\partial v_A} + \alpha y(\Sigma) = u_1(\Sigma) \quad \text{on } \Sigma, \tag{4.68}$$

$$(yx,0) = u_2(x) \quad \text{in } \Omega. \tag{4.69}$$

$$-\frac{\partial p(x,t;\mathbf{u})}{\partial t} + A[p(x,t;\mathbf{u}) = y(x,t;\mathbf{u}) - z(x,t) \quad \text{in } Q, \tag{4.70}$$

$$\frac{\partial p(\Sigma)}{\partial v_{A^*}} + \alpha(\Sigma)\,p(\Sigma) = 0 \quad \text{on } \Sigma, \tag{4.71}$$

$$p(x,T) = 0 \quad \text{in } \Omega, \tag{4.72}$$

$$p(\Sigma;\mathbf{u}) + u_1(\Sigma) - z_1(\Sigma) = 0 \quad \text{on } \Sigma, \tag{4.73}$$

$$p(x,0;\mathbf{u}) + u_2(x) - z_2(x) = 0 \quad \text{in } \Omega. \tag{4.74}$$

THEOREM 4.1,IIIB. *Given the* S_M, *hypothesis* H_M, *and the functional* $J(\mathbf{v})$ *of* PR.IIIB; *then*:

(i) *There exists onefand only one* $\mathbf{u} \in V$ *with the property that*

$$J(\mathbf{u}) \leq J(\mathbf{v}) \quad \text{for all } \mathbf{v} \in V .$$

(ii) *The boundary and initial data* $\mathbf{u}$ *is uniquely characterized by the simultaneous solution of the following system of equations*:

$$\frac{\partial y(x,t;\mathbf{u})}{\partial t} + A[y(x,t;\mathbf{u}) = f(x,t) \quad \text{in } Q , \tag{4.75}$$

$$\frac{\partial y(\Sigma)}{\partial v_A} + \alpha(\Sigma)\, y(\Sigma) = u_1(\Sigma) \quad \text{on } \Sigma , \tag{4.76}$$

$$y(x,0) = u_2(x) \quad \text{in } \Omega . \tag{4.77}$$

$$-\frac{\partial p(x,t;\mathbf{u})}{\partial t} + A[p(x,t;\mathbf{u}) = \sum_{i=1}^{v} [y(x^i,t;\mathbf{u}) - z(x^i,t)]\,\delta(x-x^i) \quad \text{in } Q , \tag{4.78}$$

$$\frac{\partial p(\Sigma)}{\partial v_{A^*}} + \alpha(\Sigma)\, p(\Sigma) = 0 \quad \text{on } \Sigma , \tag{4.79}$$

$$p(x,T) = 0 \quad \text{in } \Omega , \tag{4.80}$$

$$p(\Sigma;\mathbf{u}) + u_1(\Sigma) - z_1(\Sigma) = 0 \quad \text{on } \Sigma , \tag{4.81}$$

$$p(x,0) + u_2(x) - z_2(x) = 0 \quad \text{in } \Omega . \tag{4.82}$$

THEOREM 4.1,IIIC. *Given the system* S_M, *hypothesis* H_M, *and functional,* $J(\mathbf{v})$ *of* PR.IIIC, *then*:

(i) *There exists one and only one* $\mathbf{u} \in V$ *with the property that*

$$J(\mathbf{u}) \leq J(\mathbf{v}) \quad \text{for all } \mathbf{v} \in V .$$

(ii) *The unique solution for* $\mathbf{u}$ *is trivial and is given by*

$$u_1(\Sigma) = z_1(\Sigma) , \tag{4.83}$$

$$u_2(x) = z_2(x) . \tag{4.84}$$

A.1.1 Remarks on the Results

Having solved a variety of identification problems for the full spectrum of linear parabolic systems with spatial operator $A[\cdot]$, it is tempting, in the light of these results, to argue that these results could have been embodied in a general theorem or "maximum principle." In fact, each of the theorems given has the essence, if not the substance, of such a principle. To see this, recall that each of the problems considered had the structure

$$\inf_{\mathbf{v}\in V,\, y\in S} J(\mathbf{v}) = [a(\mathbf{v},\mathbf{v}) - 2\,\mathrm{I}(\mathbf{v}) + c]. \tag{4.85}$$

The solution to (4.85) was characterized by the system of equations

$$S: \quad \text{system evolution process,} \tag{4.86}$$

$$S^*: \quad \text{system adjoint evolution process,} \tag{4.87}$$

$$P: \quad \text{“maximum Principle,”}\ a(\mathbf{u},\mathbf{v}) - \mathrm{I}(\mathbf{v}) = 0\,. \tag{4.88}$$

Temping as it may be, the formal procedure outlined (in abstract) by (4.85)–(4.88) cannot be immediately applied as an algorithmic method for the solution of variational problems arising in the context of this study. The reason for this has been hinted at in Chapter 3, Section A.2.1. Let it be said that the procedure (4.86)–(4.88) would provide the necessary (but not necessarily sufficient) conditions for optimality. However, considerable insight would be lost by such a blind application. The stumbling block is the nature of the solutions to S and S^*, and whether these solutions play a role in the definition of $J(\mathbf{v})$.

It is possible that the system S has such a response that $J(\mathbf{v})$ has no minimum, for $\mathbf{v}\in V$. In such a situation the procedure (4.86)–(4.88) may characterize a nonexistent minimum. It is for this reason that the response $y(x, t;\mathbf{v})$ was categorized for each of the three parabolic systems considered. Further, the effect of $y(x, t; \mathbf{v})$ on the functional $J(\mathbf{v})$ was tested. In this way a solution could be guaranteed. The characterization of this solution in terms of the adjoint variable $p(x, t; \mathbf{u})$ required the establishment of existence of solutions of a particular class for $p(x, t; \mathbf{u})$. Clearly, if a solution to the adjoint system S^*, denoted by $p(x, t; \mathbf{u})$ does not exist, then the characterization in terms of $p(x, t; \mathbf{u})$ is not viable.

The solutions to the adjoint equations (4.10)–(4.12), (4.18)–(4.20), (4.49)–(4.51), (4.57)–(4.59), (4.70)–(4.72), (4.78)–(4.80) have been established [20] to be of the appropriate class.

We remarked that it is possible that S has such a response that $J(\mathbf{v})$ has no minimum. Such a case has already been considered (Chapter 3, Section A. 2.1).

In terms of the notation of this chapter the problem is:

$$\inf_{\mathbf{v}\in V,\, y\in S} J(\mathbf{v}) = \left\{ \int_{\Omega} [y(x,T;\mathbf{v}) - z(x,T)]^2\, dx\, dt \right.$$

$$+ \int_{\Sigma} [v_1(\Sigma) - z_1(\Sigma)]^2\, d\Sigma$$

$$+ \int_{\Omega} [v_2(x) - z_2(x)]^2\, d\Omega\,.$$

If we proceeded directly to the solution as implied via (4.86)–(4.88), then we would have as a characterization of $\mathbf{u} \in V$:

$$\frac{\partial y(x,t;\mathbf{u})}{\partial t} + A[y(x,t;\mathbf{u})] = f(x,t) \quad \text{in } Q\,, \tag{4.89}$$

$$y(\Sigma) = u_1(\Sigma) \quad \text{on } \Sigma\,, \tag{4.90}$$

$$y(x,0) = u_2(x) \quad \text{in } \Omega\,. \tag{4.91}$$

$$-\frac{\partial p(x,t;\mathbf{u})}{\partial t} + A[p(x,t;\mathbf{u})] = 0 \quad \text{in } Q\,, \tag{4.92}$$

$$p(\Sigma) = 0 \quad \text{on } \Sigma\,, \tag{4.93}$$

$$p(x,T) = -y(x,T;\mathbf{v}) + z(x,T) \quad \text{in } \Omega\,. \tag{4.94}$$

$$-\frac{\partial p(\Sigma;\mathbf{u})}{\partial v_{A^*}} + u_1(\Sigma) - z_1(\Sigma) = 0 \quad \text{on } \Sigma\,, \tag{4.95}$$

$$p(x,0;\mathbf{u}) + u_2(x) - z_2(x) = 0 \quad \text{in } \Omega\,. \tag{4.96}$$

It is possible that the system of equations (4.89)–(4.96) has a solution, if so, it is unique and the choice of **u** from (4.95) and (4.96) is optimal. However, under the hypothesis H_D, there are responses $y(.\,,\mathrm{T};\mathbf{u})$ which are *not* in $L^2(\Omega)$. Thus (4.89)–(4.96) *is not a maximum principle.*

From an engineering point of view, we might be persuaded to regard those inputs causing unboundedness in $y(x,T;\mathbf{u})$ as occurring with probability zero and therefore accept (4.89)–(4.96) as the solution. However, we have demonstrated only one counter example. Arguing from the same engineering point of view, we would be inclined to admit to the existence of others, and also to the possibility that these others contain "realistic" hypothesis. Thus there would be no certainty that a formal solution to the system (4.89)–(4.96) would yield the optimal **u**.

There is another approach. We could insist upon the realistic constraint that the class of inputs **u** be bounded. This added restriction to the problem complicates the mathematics, as it is necessary to employ more sophisticated methods in the proofs of some of the theorems, For completeness, such an example will be considered in Section A.3. In the context of the identification problem the assumption on boundedness of **u** corresponds to a sort of "confidence" that the "worst" possible estimates on **u** are known.

With these remarks in mind, some identification problems for hyperbolic systems are now considered.

A.2 HYPERBOLIC SYSTEMS, $V = L^2(\Sigma) \times L^2(\Omega)$

As for the parabolic systems considered in Section A.1, there are three hyperbolic systems associated with the spatial operator $A[.]$:

(I) Dirichlet boundary conditions.
(II) Neumann boundary conditions.
(III) "Mixed" boundary conditions.

For each of these systems we can pose an identification problem induced by an error functional $J(\mathbf{v})$. Provided that this error functional can be phrased in cannonical form

$$J(\mathbf{v}) = a(\mathbf{v},\mathbf{v}) - 2\mathrm{I}(\mathbf{v}) + c\,,$$

where $a(\mathbf{v},\mathbf{v})$ is a continuous bilinear form on $L^2(\Sigma)\times L^2(\Omega)$ and $\mathrm{I}(\mathbf{v})$ is a continuous linear form on $L^2(\Sigma)\times L^2(\Omega)$, then the solution of the problem is given by the solution to

$$S: \quad \text{system evolution process,} \tag{4.97}$$

$$S^*: \quad \text{system adjoint evolution process,} \tag{4.98}$$

$$P: \quad \text{"maximum principle," } a(\mathbf{u},\mathbf{v}) - \mathrm{I}(\mathbf{v}) = 0\,. \tag{4.99}$$

As in Section A.1, we shall pose the identification problem, then present its solution as a theorem. The burden of proof of the theorem will lie in the area of establishing the appropriate properties of solutions to S, S^*, and the cannonical nature of $J(\mathbf{v})$.

Again, as in Section A.1, we shall consider for each of (I), (II), and (III) identification problems induced by error functionals of the following "output" measurement processes:

$$\text{(A)}\quad z(x,t) = y(x,t) + \varepsilon(x,t) \quad \text{in } Q, \qquad z(x,t) \in L^2(Q)\,, \tag{4.100}$$

$$\text{(B)}\quad z(x^i,t) = y(x^i,t) + (x^i,t) \quad \text{in } Q, \qquad z(x^i,t) \in L^2(0,T)\,, \tag{4.101}$$

$$\text{(C)}\quad z(x,t) = 0\,. \tag{4.102}$$

The "input" measurement processes are (in each of (A), (B), and (C)):

$$z_1(\Sigma) = u_1(\Sigma)+\varepsilon_1(\Sigma) \quad \text{on } \Sigma, \qquad z(\Sigma)\in L^2(\Sigma), \tag{4.103}$$

$$z_2(x,0) = u_2(x)+\varepsilon_2(x) \quad \text{in } \Omega, \tag{4.104}$$

$$z_3(x,0) = u_3(x)+\varepsilon_3(x) \quad \text{in } \Omega. \tag{4.105}$$

Define the following error functionals:

PR.A.

$$J(\mathbf{v}) = \int_Q [y(x,t;\mathbf{v})-z(x,t)]^2\,dx\,dt + \int_\Sigma [v_1(\Sigma)-z_1(\Sigma)]^2\,d\Sigma + \int_\Omega \{[v_2(x)-z_2(x)]^2+[v_3(x)-z_3(x)]^2\}\,dx \tag{4.106}$$

PR.B.

$$J(\mathbf{v}) = \sum_{i=1}^{\nu}\int_0^T [y(x^i,t;\mathbf{v})-z(x^i,t)]^2\,dx\,dt + \int_\Sigma [v_1(\Sigma)-z_1(\Sigma)]^2\,d\Sigma + \int_\Omega \{[v_2(x)-z_2(x)]^2+[v_3(x)-z_3(x)]^2\,dx. \tag{4.107}$$

PR.C.

$$J(\mathbf{v}) = \int_\Sigma [v_1(\Sigma)-z_1(\Sigma)]^2\,d\Sigma + \int_\Omega \{[v_2(x)-z_2(x)]^2+[v_3(x)-z_3(x)]^2\}\,dx. \tag{4.108}$$

In each of PR.A, PR.B, and PR,C, $y(.,\cdot;\mathbf{v})$ is a trajectory of a hyperbolic system (I), (II), or (III). Consider the following identification problems and their solution:

PR.I. *Dirichlet Boundary Conditions*

System S_D.

$$\frac{\partial^2 y(x,t)}{\partial t^2} + A[y(x,t)] = f(x,t) \quad \text{in } Q, \tag{4.109}$$

$$y(\Sigma) = v_1(\Sigma) \quad \text{on } \Sigma. \tag{4.110}$$

$$y(x,0) = v_2(x) \quad \text{in } \Omega\,, \tag{4.111}$$

$$\frac{\partial y(x,0)}{\partial t} = v_3(x) \quad \text{in } \Omega\,. \tag{4.112}$$

Hypothesis H_D.

$$f(\cdot,\cdot) \in L^2(Q): \quad \int_Q |f(x,t)|^2\,dx\,dt < \infty\,,$$

$$\left.\begin{array}{ll} v_1(.) \in L^2(\Sigma): & \displaystyle\int_\Sigma |v_1(\Sigma)|^2\,d\Sigma < \infty \\ v_2(\cdot) \in L^2(\Omega): & \displaystyle\int_\Omega |v_2(x)|^2\,dx < \infty \\ v_3(\cdot) \in L^2(\Omega): & \displaystyle\int_\Omega |v_3(x)|^2\,dx < \infty \end{array}\right\} = V\,.$$

Additional hypothesis on $A[\cdot]$.

$$a_{ij}(x,t) = a_{ji}(x,t), \qquad i,j = 1,2,3,\ldots,r\,.$$

Remark. The symmetry on the coefficients of the differential operator $A[\cdot]$ was not neccesary in the parabolic case but is essential here. For the system S_D and hypothesis H_D we give the following results as theorems:

THEOREM 4.2,IA. *Given the system S_D, hypothesis H_D, and the functional $J(\mathbf{v})$ of* PR.A, *then:*

(i) *There exists one and only one $\mathbf{u} \in V$ with the property that*

$$J(\mathbf{u}) \leq J(\mathbf{v}) \quad \text{for all } \mathbf{v} \in V\,.$$

(We mean by $\mathbf{u}$ the vector $[u_1(\Sigma)u_2(x)u_3(x)]^T$.)

(ii) *The boundary and initial data $\mathbf{u}$ are uniquely characterized by the simultaneous solution of the following system of equations:*

$$\frac{\partial^2 y(x,t;\mathbf{u})}{\partial t^2} + A[y(x,t;\mathbf{u})] = f(x,t) \quad \text{in } Q\,, \tag{4.113}$$

$$y(\Sigma) = u_1(\Sigma) \quad \text{on } \Sigma\,, \tag{4.114}$$

$$y(x,0) = u_2(x) \quad \text{in } \Omega\,, \tag{4.115}$$

$$\frac{\partial y(x,0)}{\partial t} = u_3(x) \quad \text{in } \Omega\,. \tag{4.116}$$

$$\frac{\partial^2 p(x,t;\mathbf{u})}{\partial t^2} + A[p(x,t;\mathbf{u})] = y(x,t;\mathbf{u}) - z(x,t) \quad \text{in } Q, \tag{4.117}$$

$$p(\Sigma) = 0 \quad \text{on } \Sigma, \tag{4.118}$$

$$p(x,T) = 0 \quad \text{in } \Omega, \tag{4.119}$$

$$\frac{\partial p(x,T)}{\partial t} = 0 \quad \text{in } \Omega, \tag{4.120}$$

$$-\frac{\partial p(\Sigma;\mathbf{u})}{\partial \nu_{A^*}} + u_1(\Sigma) - z_1(\Sigma) = 0 \quad \text{on } \Sigma, \tag{4.121}$$

$$p(x,0;\mathbf{u}) + u_2(x) - z_2(x) = 0 \quad \text{in } \Omega, \tag{4.122}$$

$$-\frac{\partial p(x,0;\mathbf{u})}{\partial t} + u_3(x) - z_3(x) = 0 \quad \text{in } \Omega. \tag{4.123}$$

THEOREM 4.2,IB. *Given the system* S_D, *hypothesis* H_D *and functional* $J(\mathbf{v})$ *of* PR.B, *then*:

(i) *There exists one and only one* $\mathbf{u} \in V$ *with the property that*

$$J(\mathbf{u}) \leq J(\mathbf{v}) \quad \text{for all } \mathbf{v} \in V.$$

(ii) *The boundary and initial data* $\mathbf{u}$ *is uniquely characterized by the simultaneous solution of the following system of equations*:

$$\frac{\partial^2 y(x,t;\mathbf{u})}{\partial t^2} + A[y(x,t;\mathbf{u})] = f(x,t) \quad \text{in } Q, \tag{4.124}$$

$$y(\Sigma) = u_1(\Sigma) \quad \text{on } \Sigma, \tag{4.125}$$

$$y(x,0) = u_2(x) \quad \text{in } \Omega, \tag{4.126}$$

$$\frac{\partial y(x,0)}{\partial t} = u_3(x) \quad \text{in } \Omega. \tag{4.127}$$

$$\frac{\partial^2 p(x,t;\mathbf{u})}{\partial t^2} + A[p(x,t;\mathbf{u}) = \sum_{i=1}^{\nu} [y(x^i,t;\mathbf{u}) - z(x^i,t)]\,\delta(x-x^i) \quad \text{in } Q, \tag{4.128}$$

$$p(\Sigma;\mathbf{u}) = 0 \quad \text{on } \Sigma, \tag{4.129}$$

$$p(x,T) = 0 \quad \text{in } \Omega, \tag{4.130}$$

$$\frac{\partial p(x,T)}{\partial t}=0 \quad \text{in } \Omega . \tag{4.131}$$

$$-\frac{\partial p(\Sigma;\mathbf{u})}{\partial \nu_{A^*}}+u_1(\Sigma)-z_1(\Sigma)=0 \quad \text{on } \Sigma , \tag{4.132}$$

$$p(x,0;\mathbf{u})+u_2(x)-z_2(x)=0 \quad \text{in } \Omega , \tag{4.133}$$

$$-\frac{\partial p(x,0;\mathbf{u})}{\partial t}+u_3(x)-z_3(x)=0 \quad \text{in } \Omega . \tag{4.134}$$

Remark.

$\frac{\partial p}{\partial \nu_{A^*}}=\frac{\partial p}{\partial \nu_A}$ for hyperbolic systems under the hypothesis on $A[\cdot]=A^*[\cdot]$

THEOREM IV.2,IC. *Given the system S_N, hypothesis H_N, and the functional $J(\mathbf{v})$ of* PR.C, *then*:

(i) *There exists one and only one $\mathbf{u}\in V$ with the property that*:

$$J(\mathbf{u})\le J(\mathbf{v}) \quad \text{for all } \mathbf{v}\in V .$$

(ii) *The unique solution for* $\mathbf{u}$ *is trivial and is given by*

$$u_1(\Sigma)=z_1(\Sigma) \quad \text{on } \Sigma , \tag{4.135}$$
$$u_2(x)=z_2(x) \quad \text{in } \Omega , \tag{4.136}$$

$$u_3(x)=z_3(x) \quad \text{in } \Omega . \tag{4.137}$$

Remark on the Proofs of Theorems 4.2,IA, 4.2,IB, *and* 4.2,IC. The proofs of these theorems carry through *exactly* as for Theorems 4.1,IA, 4.1,IB, and 4.1,IC. The additional hypothesis on $A[\cdot]$ given in H_D over and above the hypothesis on $A[\cdot]$ for parabolic systems (namely, that $A[\cdot]=A^*[\cdot]$) allows this exact equivalence. The ramification of this additional hypothesis is that the responses for $y(x,t;\mathbf{u})$ and $p(x,t;\mathbf{u})$ are of the same classes as in the corresponding parabolic systems.

PR.II. *Neumann Boundary Conditions*

System S_N.

$$\frac{\partial^2 y(x,t)}{\partial t^2}+A[y(x,t)]=f(x,t) \quad \text{in } Q , \tag{4.138}$$

$$\frac{\partial y(\Sigma)}{\partial \nu_A}=v_1(\Sigma) \quad \text{on } \Sigma , \tag{4.139}$$

$$y(x,0) = v_2(x) \quad \text{in } \Omega, \tag{4.140}$$

$$\frac{\partial y(x,0)}{\partial t} = v_3(x) \quad \text{in } \Omega. \tag{4.141}$$

Hypothesis H_N.

$$f(\cdot,\cdot) \in L(Q): \quad \int_Q |f(x,t)|^2 \, dx\, dt < \infty,$$

$$\left.\begin{aligned} v_1(\cdot) \in L^2(\Sigma): &\quad \int_\Sigma |v_1(\Sigma)|^2 \, d\Sigma < \infty \\ v_2(\cdot) \in L^2(\Omega): &\quad \int_\Omega |v_2(x)|^2 \, dx < \infty \\ v_3(\cdot) \in L^2(\Omega): &\quad \int_\Omega |v_3(x)|^2 \, dx < \infty \end{aligned}\right\} = V.$$

Additional Hypothesis on $A[.]$.

$$a_{ij}(x,t) = a_{ji}(x,t), \qquad i,j = 1,2,3,\ldots,r.$$

THEOREM IV.2,IIA. *Given the system S_N, hypothesis H_N, and the functional $J(\mathbf{v})$ of* PR.A, *then*:

(i) *There exists one and only one* $\mathbf{u} \in V$ *with the property that*

$$J(\mathbf{u}) \le J(\mathbf{v}) \quad \text{for all } \mathbf{v} \in V.$$

(ii) *The boundary and initial data* $\mathbf{u}$ *is uniquely characterized by the simultaneous solution of the following system of equations*:

$$\frac{\partial^2 y(x,t;\mathbf{u})}{\partial t^2} + A[y(x,t;\mathbf{u})] = f(x,t) \quad \text{in } Q, \tag{4.142}$$

$$\frac{\partial y(\Sigma)}{\partial \nu_A} = u_1(\Sigma) \quad \text{on } \Sigma, \tag{4.143}$$

$$y(x,0) = u_2(x) \quad \text{in } \Omega, \tag{4.144}$$

$$\frac{\partial y(x,0)}{\partial t} = u_3(x) \quad \text{in } \Omega. \tag{4.145}$$

$$\frac{\partial^2 p(x,t;\mathbf{u})}{\partial t^2} + A[p(x,t;\mathbf{u})] = y(x,t;\mathbf{u}) - z(x,t) \quad \text{in } Q, \tag{4.146}$$

$$\left.\begin{array}{lll} v_1(\cdot)\in L^2(\Sigma): & \displaystyle\int_\Sigma |v_1(\Sigma)|^2\,d\Sigma & <\infty \\ v_2(\cdot)\in L^2(\Omega): & \displaystyle\int_\Omega |v_2(x)|^2\,d\Omega & <\infty \\ v_3(\cdot)\in L^2(\Omega): & \displaystyle\int_\Omega |v_3(x)|^2\,d\Omega & <\infty \end{array}\right\} = V,$$

$$\alpha(\Sigma)\geq 0 \quad \text{on } \Sigma.$$

Additional Hypothesis on $A[.]$.

$$a_{ij}(x,t) = a_{ji}(x,t), \qquad i,j = 1,2,3,\ldots,r.$$

THEOREM IV.2,IIIA. *Given the system* S_M, *hypothesis* H_M, *and functional* $J(\mathbf{v})$ *of* PR.A, *then*:

(i) *There exists one and only one* $\mathbf{u}\in V$ *with the property that*:

$$J(\mathbf{u}) \leq J(\mathbf{v}), \quad \text{for all } \mathbf{v}\in V.$$

(ii) *The boundary and initial data* $\mathbf{u}$ *is uniquely determined by the simultaneous solution of the following system of equations*:

$$\frac{\partial^2 y(x,t;\mathbf{u})}{\partial t^2} + A[y(x,t;\mathbf{u})] = f(x,t) \quad \text{in } Q, \tag{4.171}$$

$$\frac{\partial y(\Sigma)}{\partial v_A} + \alpha(\Sigma)\,y(\Sigma) = u_1(\Sigma) \quad \text{on } \Sigma, \tag{4.172}$$

$$y(x,0) = u_2(x) \quad \text{in } \Omega, \tag{4.173}$$

$$\frac{\partial y(x,0)}{\partial t} = u_3(x) \quad \text{in } \Omega. \tag{4.174}$$

$$\frac{\partial^2 p(x,t;\mathbf{u})}{\partial t^2} + A[p(x,t;\mathbf{u})] = y(x,t;\mathbf{u}) - z(x,t) \tag{4.175}$$

$$\frac{\partial p(\Sigma;\mathbf{u})}{\partial v_{A^*}} + \alpha(\Sigma)\,p(\Sigma;\mathbf{u}) = 0 \quad \text{on } \Sigma, \tag{4.176}$$

$$p(x,T;\mathbf{u}) = 0 \quad \text{in } \Omega, \tag{4.177}$$

$$\frac{\partial p(x,T;\mathbf{u})}{\partial t} = 0 \quad \text{in } \Omega, \tag{4.178}$$

$$p(\Sigma;\mathbf{u})+u_1(\Sigma)-z_1(\Sigma)=0 \quad \text{on } \Sigma\,, \tag{4.161}$$

$$p(x,0;\mathbf{u})+u_2(x)-z_2(x)=0 \quad \text{in } \Omega\,, \tag{4.162}$$

$$-\frac{\partial p(x,0;\mathbf{u})}{\partial t}+u_3(x)-z_3(x)=0 \quad \text{in } \Omega\,. \tag{4.163}$$

THEOREM IV.2,IIC. *Given the system S_N, hypothesis H_N, and the functional $J(\mathbf{v})$ of* PR.C, *then*:

(i) *There exists one and only one $\mathbf{u}\in V$ with the property that*:

$$J(\mathbf{u})\le J(\mathbf{v}) \quad \text{for all } \mathbf{v}\in V\,.$$

(ii) *The unique solution for* $\mathbf{u}$ *is trivial and is given by*

$$u_1(\Sigma)=z_1(\Sigma) \quad \text{on } \Sigma\,, \tag{4.164}$$

$$u_2(x)=z_2(x) \quad \text{in } \Omega\,, \tag{4.165}$$

$$u_3(x)=z_3(x) \quad \text{in } \Omega\,, \tag{4.166}$$

Remark on the Proofs of Theorems IV.2,IIA, IV.2,IIB, *and* IV.2,IIC. The proofs of these theorems carry through exactly as for Theorems IV.1,IIA, IV.1,IIB, and IV.1,IIC, the additional hypothesis on the symmetry of $A[\cdot]$ having been added.

PR.III. *Mixed Boundary Conditions*

System S_M.

$$\frac{\partial^2 y(x,t)}{\partial t^2}+A[y(x,t)]=f(x,t) \quad \text{in } Q\,, \tag{4.167}$$

$$\frac{\partial y(\Sigma)}{\partial \nu_A}+\alpha(\Sigma)\,y(\Sigma)=v_1(\Sigma) \quad \text{on } \Sigma\,, \tag{4.168}$$

$$y(x,0)=v_2(x) \quad \text{in } \Omega\,, \tag{4.169}$$

$$\frac{\partial y(x,0)}{\partial t}=v_3(x) \quad \text{in } \Omega\,. \tag{4.170}$$

Hypothesis H_M.

$$f(\cdot,\cdot)\in L^2(Q):\ \int_Q |f(x,t)|^2\,dx\,dt<\infty\,,$$

$$\left.\begin{aligned} v_1(\cdot)\in L^2(\Sigma)&: \quad \int_\Sigma |v_1(\Sigma)|^2\,d\Sigma < \infty \\ v_2(\cdot)\in L^2(\Omega)&: \quad \int_\Omega |v_2(x)|^2\,d\Omega < \infty \\ v_3(\cdot)\in L^2(\Omega)&: \quad \int_\Omega |v_3(x)|^2\,d\Omega < \infty \end{aligned}\right\} = V,$$

$$\alpha(\Sigma) \geq 0 \quad \text{on } \Sigma.$$

Additional Hypothesis on $A[.]$.

$$a_{ij}(x,t) = a_{ji}(x,t), \qquad i,j = 1,2,3,\ldots,r.$$

THEOREM IV.2,IIIA. *Given the system* S_M, *hypothesis* H_M, *and functional* $J(\mathbf{v})$ *of* PR.A, *then*:

(i) *There exists one and only one* $\mathbf{u}\in V$ *with the property that*:

$$J(\mathbf{u}) \leq J(\mathbf{v}), \quad \text{for all } \mathbf{v}\in V.$$

(ii) *The boundary and initial data* $\mathbf{u}$ *is uniquely determined by the simultaneous solution of the following system of equations*:

$$\frac{\partial^2 y(x,t;\mathbf{u})}{\partial t^2} + A[y(x,t;\mathbf{u})] = f(x,t) \quad \text{in } Q, \tag{4.171}$$

$$\frac{\partial y(\Sigma)}{\partial \nu_A} + \alpha(\Sigma)\,y(\Sigma) = u_1(\Sigma) \quad \text{on } \Sigma, \tag{4.172}$$

$$y(x,0) = u_2(x) \quad \text{in } \Omega, \tag{4.173}$$

$$\frac{\partial y(x,0)}{\partial t} = u_3(x) \quad \text{in } \Omega. \tag{4.174}$$

$$\frac{\partial^2 p(x,t;\mathbf{u})}{\partial t^2} + A[p(x,t;\mathbf{u})] = y(x,t;\mathbf{u}) - z(x,t) \tag{4.175}$$

$$\frac{\partial p(\Sigma;\mathbf{u})}{\partial \nu_{A^*}} + \alpha(\Sigma)\,p(\Sigma;\mathbf{u}) = 0 \quad \text{on } \Sigma, \tag{4.176}$$

$$p(x,T;\mathbf{u}) = 0 \quad \text{in } \Omega, \tag{4.177}$$

$$\frac{\partial p(x,T;\mathbf{u})}{\partial t} = 0 \quad \text{in } \Omega, \tag{4.178}$$

$$p(\Sigma;\mathbf{u})+u_1(\Sigma)-z_1(\Sigma)=0 \quad \text{on } \Sigma, \tag{4.179}$$

$$p(x,0;\mathbf{u})+u_2(x)-z_2(x)=0 \quad \text{in } \Omega, \tag{4.180}$$

$$-\frac{\partial p(x,0;\mathbf{u})}{\partial t}+u_3(x)-z_3(x)=0 \quad \text{in } \Omega. \tag{4.181}$$

THEOREM VI.2,IIIB. *Given the system S_M, hypothesis H_M, and the functional $J(\mathbf{v})$ of* PR.B, *then*:

(i) *There exists one and only one $\mathbf{u}\in V$ with the property that*:

$$J(\mathbf{u}) \le J(\mathbf{v}), \quad \text{for all } \mathbf{v}\in V.$$

(ii) *The boundary and initial data* $\mathbf{u}$ *is uniquely determined by the simultaneous solution of the following system of equations*:

$$\frac{\partial^2 y(x,t;\mathbf{u})}{\partial t^2}+A[y(x,t;\mathbf{u})]=f(x,t) \quad \text{in } Q, \tag{4.182}$$

$$\frac{\partial y(\Sigma)}{\partial \nu_A}+\alpha(\Sigma)\,y(\Sigma)=u_1(\Sigma) \quad \text{on } \Sigma, \tag{4.183}$$

$$y(x,0)=u_2(x) \quad \text{in } \Omega, \tag{4.184}$$

$$\frac{\partial y(x,0)}{\partial t}=u_3(x) \quad \text{in } \Omega. \tag{4.185}$$

$$\frac{\partial^2 p(x,t;\mathbf{u})}{\partial t^2}+A[p(x,t;\mathbf{u})]=\sum_{i=1}^{\nu}[y(x^i,t;\mathbf{u})-z(x^i,t)]\,\delta(x-x^i)$$
$$\text{in } Q, \tag{4.186}$$

$$\frac{\partial p(\Sigma;\mathbf{u})}{\partial \nu_{A^*}}+\alpha(\Sigma)\,p(\Sigma;\mathbf{u})=0 \quad \text{on } \Sigma, \tag{4.187}$$

$$p(x,T;\mathbf{u})=0 \quad \text{in } \Omega, \tag{4.188}$$

$$\frac{\partial p(x,T;\mathbf{u})}{\partial t}=0 \quad \text{in } \Omega, \tag{4.189}$$

$$p(\Sigma;\mathbf{u})+u_1(\Sigma)-z_1(\Sigma)=0 \quad \text{on } \Sigma, \tag{4.190}$$

$$p(x,0;\mathbf{u})+u_2(x)-z_2(x)=0 \quad \text{in } \Omega, \tag{4.191}$$

$$-\frac{\partial p(x,0;\mathbf{u})}{\partial t}+u_3(x)-z_3(x)=0 \quad \text{in } \Omega. \tag{4.192}$$

THEOREM IV.2,IIIC. *Given the system* S_M, *hypothesis* H_M, *and functional* $J(\mathbf{v})$ *of* PR.C, *then*:

(i) *There exists one and only one* $\mathbf{u} \in V$ *with the property that*:

$$J(\mathbf{u}) \leq J(\mathbf{v}) \quad \text{for all } \mathbf{v} \in V .$$

(ii) *The unique solution for* $\mathbf{u}$ *is trivial and is given by*

$$u_1(\Sigma) = z_1(\Sigma) \quad \text{on } \Sigma , \tag{4.193}$$

$$u_2(x) = z_2(x) \quad \text{in } \Omega , \tag{4.194}$$

$$u_3(x) = z_3(x) \quad \text{in } \Omega , \tag{4.195}$$

Remark on the Proof sof Theorems IV.2,IIIA, IV.2,IIIB, *and* IV.2,IIIC. The proofs of these theorems carry through exactly as for Theorems IV.2,IIIA, IV.2,IIIB, and IV.2,IIIC, the hypothesis regarding the summetry of $A[\cdot]$ having been added.

We have now exhausted (and have been exhausted by) the combinations of identification problems for parabolic and hyperbolic systems in which

(a) the admissible set $V = L^2(\Sigma) \times L^2(\Omega)$,

(b) the error functionals are of type PR.A, PR.B, or PR.C.

Clearly, many other functionals could be constructed which obey the hypothesis that

$$J(\mathbf{v}) = a(\mathbf{v}, \mathbf{v}) - 2\mathrm{I}(\mathbf{v}) + c ,$$

where $a(\mathbf{v}, \mathbf{v})$ and $\mathrm{I}(\mathbf{v})$ are continuous linear and bilinear forms on $L^2(\Sigma) \times L^2(\Omega)$. We believe that the technique is sufficiently clear that such functionals could be handled in the framework of this chapter.

The case where $V \subset L^2(\Sigma) \times L^2(\Omega)$ can also be studied within this framework, provided V is closed and convex. As an illustrative example, we now consider such a case.

A.3 AN EXAMPLE OF CONSTRAINED STATE IDENTIFICATION

As we mentioned in the closing paragraph of Section A.1, it is reasonable to pose the following identification problem for a given system S:

P. Given a system S, input and output measurements I and O, respectively where the input measurements are known with a specified confidence, denoted by C, Then obtain a refinement of these measurements, R, along a trajectory of S.

Problem P corresponds to a physically appealing notion that the error processes ε associated with the boundary and initial data measurements is bounded by known quantities $\boldsymbol{\lambda}_1$ and $\boldsymbol{\lambda}_2$:

$$\boldsymbol{\lambda}_1 \leq \boldsymbol{\varepsilon} \leq \boldsymbol{\lambda}_2 .$$

P. *Identification Problem for Bounded Input* $\mathbf{v} \in V$, $L^2(\Sigma)\varepsilon L^2(\Omega)$

System S_D.

$$\frac{\partial y(x,t)}{\partial t} + A[y(x,t)] = f(x,t) \quad \text{in } Q, \tag{4.196}$$

$$y(\Sigma) = u_1(\Sigma) \quad \text{on } \Sigma, \tag{4.197}$$

$$y(x,0) = u_2(\Sigma) \quad \text{in } \Omega. \tag{4.198}$$

Hypothesis H_D.

$$f(\cdot,\cdot)L^2(Q),$$

$$\left.\begin{array}{l} v_1(\cdot)\varepsilon L^2(\Sigma) \quad \text{and} \quad \lambda_1{}^L \leq v_1(\Sigma) \leq \lambda_1{}^U \\ v_2(\cdot)\varepsilon L^2(\Omega) \quad \text{and} \quad \lambda_2{}^L \leq v_2(x) \leq \lambda_2{}^U \end{array}\right\} = V.$$

Then we have the following theorem:

THEOREM 4.3. *Given the system* S_D, *hypothesis*H_D, *and the functional* $J(\mathbf{v})$ *of* PR.A, *then*:

(i) *There is one and only one* $\mathbf{u} \in V$ *with the property that*:

$$J(\mathbf{u}) \leq J(\mathbf{v}) \quad \text{for all } \mathbf{v} \in V.$$

(ii) *The boundary and initial data* $\mathbf{u}$ *is uniquely characterized by the simultaneous solution of the following system of equations*:

$$\frac{\partial y(x,t;\mathbf{u})}{\partial t} + A[y(x,t;\mathbf{u})] = f(x,t) \quad \text{in } Q, \tag{4.199}$$

$$y(\Sigma) = u_1(\Sigma) \quad \text{on } \Sigma, \tag{4.200}$$

$$y(x,0) = u_2(x) \quad \text{in } \Omega. \tag{4.201}$$

$$\frac{\partial p(x,t;\mathbf{u})}{\partial t} + A[p(x,t;\mathbf{u})] = y(x,t;\mathbf{u}) - z(x,t) \quad \text{in } Q, \tag{4.202}$$

$$p(\Sigma) = 0 \quad \text{on } \Sigma, \tag{2.203}$$

$$p(x,T) = 0 \quad \text{in } \Omega. \tag{2.204}$$

$$\left[-\frac{\partial p(\Sigma;\mathbf{u})}{\partial \nu_{A^*}}+u_1(\Sigma)-z_1(\Sigma)\right][\lambda_1-u_1(\Sigma)]\geq 0,$$

$$\lambda_1{}^L\leq\lambda_1\leq\lambda_1{}^U,\quad \text{on }\Sigma, \tag{4.205}$$

$$[p(x,0;\mathbf{u})+u_2(x)-z_2(x)][\lambda_2-u_2(x)]\geq 0,$$

$$\lambda_2{}^L\leq\lambda_2\leq\lambda_2^U,\quad \text{in }\Omega. \tag{4.206}$$

In the proof of Theorem 4.3 it is necessary to invoke the following lemma due to Lebesque:

LEMMA 4.7. *Given a measurable function $\Psi(x,t)$ defined on Q. If 0_j is an elemental volume of Q at M_0 with measure $\mu(O_j)$ and*

$$\lim_{j\to\infty} 0_j\to 0,$$

then if

$$\frac{1}{\mu(0_j)}\int_{0_j}\Psi(x,t)\,dx\,dt\geq 0,$$

$$\lim_{j\to\infty}\frac{1}{\mu(0_j)}\int_{0_j}\Psi(x,t)\,dx\,dt=\Psi(M_0)\geq 0$$

almost everywhere in Q.

Remark on Lemma 4.7. Note the weak hypothesis on $\Psi(x,t)$. In particular, $\Psi(x,t)$ is not necessarily continuous in Q.

PROOF OF THEOREM 4.3. Invoke Lemma 4.4 and write $J(\mathbf{v})$ of PR.A as:

$$J(\mathbf{v})=a(\mathbf{v},\mathbf{v})-2\mathrm{I}(\mathbf{v})+c\,;$$

in addition,

$$V=\{\mathbf{v}\colon\ \mathbf{v}\in L^2(\Sigma)\times L^2(\Omega);\boldsymbol{\lambda}^L\leq\mathbf{v}\leq\boldsymbol{\lambda}^U\}.$$

Evidently,

(a) V is closed,

(b) V is convex.*

* for $k_1,\,k_2\in V,\quad 0\leq\mu\leq 1,$
$\mathbf{v}_\mu=\mu k_1+(1-\mu)\,k_2\,\varepsilon\,V,\qquad v_\mu\in L^2(\Sigma)\times L^2(\Omega)\,;\qquad \lambda^L\leq v_\mu\leq\lambda^U.$

Thus the hypothesis of Theorem 2.4 is satisfied and so we have immediately that

(i) *There exists one and only one* $\mathbf{u} \in V$ *such that*

$$J(\mathbf{u}) \leq J(\mathbf{v}) \quad \text{for all } \mathbf{v} \in V .$$

Moreover, $\mathbf{u}$ is characterized by

$$a(\mathbf{u}, \mathbf{v}-\mathbf{v}) \geq \mathrm{I}(\mathbf{v}, -\mathbf{u}) \quad \text{for all } \mathbf{v} \in V .$$

Using the definitions (4.26) and (4.27), (4.208) is equivalent to:

$$\int_Q [y(x,t;\mathbf{u}) - z(x,t)][y(x,t;\mathbf{v}) - y(x,t;\mathbf{u})]\,dx\,dt$$
$$+ \int_\Sigma [u_1(\Sigma) - z_1(\Sigma)][v_1(\Sigma) - u_1(\Sigma)]\,d\Sigma$$
$$+ \int_\Omega [u_2(x) - z_2(x)][v_2(x) - u_2(x)]\,dx = 0 . \tag{4.209}$$

Introduce the system adjoint to S_D:

$$\frac{\partial p(x,t;\mathbf{u})}{\partial t} + A[p(x,t;\mathbf{u})] = y(x,t;\mathbf{u}) - z(x,t) \quad \text{in } Q , \tag{4.210}$$

$$p(\Sigma;\mathbf{u}) = 0 \quad \text{on } \Sigma , \tag{4.211}$$

$$p(x,T;\mathbf{u}) = 0 \quad \text{in } \Omega . \tag{4.212}$$

As we saw in the proof of Theorem 4.1,IA, the adjoint system can be arranged to give an alternative representation to (4.209). By multiplying (4.210) by $[y(x,t;\mathbf{v}) - y(x,t;\mathbf{u})]$ and then integrating the result over Q, it is seen that substitution of (4.209) into the resulting expression yields on application of Green's theorem, the result

$$\int_\Sigma \left[-\frac{\partial p(\Sigma;\mathbf{u})}{\partial v_{A^*}} + u_1(\Sigma) - z_1(\Sigma) \right] [v_1(\Sigma) - u_1(\Sigma)]\,d\Sigma \geq 0 , \tag{4.213}$$

$$\int_\Omega [p(x,0;\mathbf{u}) + u_2(x) - z_2(x)][v_2(x) - z_2(x)]\,dx \geq 0 . \tag{4.214}$$

Assertion. Expressions (4.213) and (4.214) are equivalent to

$$\left[-\frac{\partial p(\Sigma;\mathbf{u})}{\partial v_{A^*}} + u_1(\Sigma) - z_1(\Sigma) \right] [\lambda_1 - u_1(\Sigma)] \geq 0 , \qquad \lambda_1{}^L \leq \lambda_1 \leq \lambda_1{}^U , \tag{4.215}$$

$$[p(x,0;\mathbf{u}) + u_2(x) - z_2(x)][\lambda_2 - u_2(x)] \geq 0 , \qquad \lambda_1{}^L \leq \lambda_2 \leq \lambda_2{}^U , \tag{4.216}$$

PROOF. Evidently, (4.215) and (4.216) imply (4.213) and (4.214), respectively. Simply replace λ_1 by $v_1(\Sigma)$ and λ_2 by $v_2(x)$ and integrate appropriately.

However, it is not obvious that (4.213) and (4.214) imply (4.215) and (4.216), respectively. If the integrands of (4.213) and (4.214) were continuous, then the implication would be obvious. However, the hypothesis H_D does not permit such a restrictive assumption.

We shall obtain the desired result by the application of Lemma 4.7.

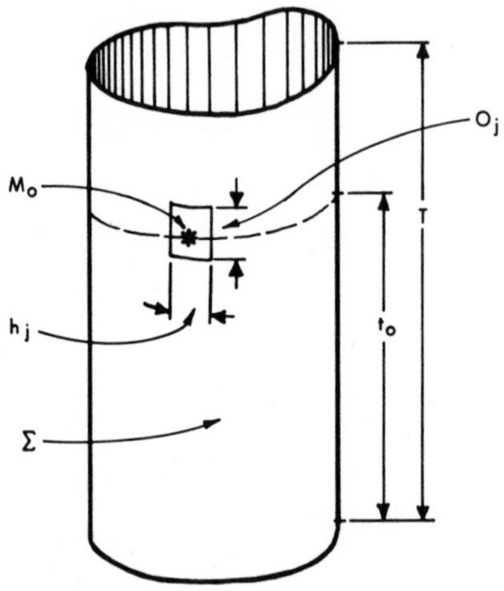

Figure 4.1. Position of the elemental area *Aj* on the surface Σ.

Consider Figure 4.1. Σ is the surface of the cyclindrical sheath to the volume Q. Ω is the "bottom" surface of this volume Q. The boundary or perimeter of Ω is Γ. Let M_0 be a point on the surface Σ with coordinates (Γ_0, t_0). Let M_0 be contained in an elemental area A_j with sides h_j and 0_j. Let

$$\lim_{j\to\infty} 0_j \to 0, \qquad \lim_{j\to\infty} h_j \to 0.$$

Consider as a "test function" $v_j(\Sigma)$, defined as follows:

$$v_j(\Sigma) = \begin{cases} \lambda_1 & \text{inside } A_j; \quad \lambda_1{}^L \le \lambda_1 \le \lambda_1{}^U \\ u_1(\Sigma) & \text{outside } A_j; \quad u_1 \in V \end{cases} \tag{4.217}$$

Evidently, $v_j \in V$. Thus, using (4.213), we obtain:

$$\int_{A_j}\left[-\frac{\partial p}{\partial v_{A^*}}+u_1-z_1\right][\lambda-u_1]\,d\mu$$

$$+\int_{\Sigma-A_j}\left[-\frac{\partial p}{\partial v_{A^*}}+u_1-z_1\right][u_1-u_1]\,d\mu \geq 0, \tag{4.218}$$

where $d\mu$ is a differential area for the surface Σ. Since the second term in (4.218) is zero, we have, on multiplication by $1/\mu(A_j)$,

$$\frac{1}{\mu(A_j)}\int_{A_j}\left[-\frac{\partial p}{\partial v_{A^*}}+u_1-z_1\right][\lambda-u_1]\,d\mu \geq 0 \tag{4.219}$$

since $\mu(A_j) \geq 0$. $\mu(A_j)$ is the "measure" of A_j (for engineering purposes, its area).

We now invoke Lemma 4.7 and obtain that

$$\left[-\frac{\partial p(\Gamma_0,t_0)}{\partial v_{A^*}}+u_1(\Gamma_0,t_0)-z_1(\Gamma_0,t_0)\right][\lambda_1-u_1(\Gamma_0,t_0)] \geq 0. \tag{4.220}$$

But (4.220) is true for almost every Γ_0, t_0 in Σ. Arguing similarily in Ω, the assertion follows.

Thus Part (ii) of the theorem follows, in virtue of the uniqueness of the solution to (4.199)–(4.204).

The problem giving rise to these equations is somewhat contrived. However, in cases where the measurement error contains significant noise "bursts," the computational burden associated with this formulation may be worthwhile.

A.4 CONCLUDING REMARKS

The state identification problem for parabolic and hyperbolic systems associated with the operator $A[\cdot]$ has been solved (i.e., a characterization of the solution has been obtained) for some rather simple quadratic error functionals. The technique of solutions was demonstrated via several complete proofs concerning existence and uniqueness of optimal $\mathbf{v} \in V$, for V bounded and unbounded. The technique is valid for any choice of quadratic error functional, as it filters out those functionals which are non-well set.

The proof of Theorem 4.3 was of special importance, since the characterization of the extremal to the identification problem was in terms of a pair of inequations. These inequations were shown to be a *valid and equivalent representation* of thr condition of optimality:

$$a(\mathbf{u}, \mathbf{v}-\mathbf{u}) \geq \mathrm{I}(\mathbf{v}-\mathbf{u})\,.$$

To our knowledge, this constitutes the only treatment of what could be called the "constrained estimate" state identification problem. Though of limited practical importance, our results for this case are theoretically interesting, paralleling results in optimal control theory for problems with control constraints.

CHAPTER 5

SOLUTION OF THE STATE IDENTIFICATION PROBLEM

A.0 INTRODUCTION

In Chapter 4, the solution to the state identification problem associated with the distributed system S was characterized by a set of simultaneous equations, denoted here by E.

In this chapter we consider two methods which construct a solution to E. We remark that the construction ot solution to E is not straightforward, owing to the split end conditions in time contained therein.

Specifically, a Riccati-like decoupling of the two point "time boundary" value problem is considered in Section A.1.1. In Section A.1.2 we solve E by solving directly the variational problem leading to E. This solution is accomplished by hill climbing (descending) on the error functional $J(\mathbf{v})$. Directions of search on the hill are conjugate.

Both methods for solving E lead to the consideration of a set of partial differential equations the solution of which must be approximated, since they defy (in general) an analytic representation of any immediate numerical utility. Section A.2 deals with the application of the method of Galerkin to the results of Section A.1, Chapter 5 concludes with the consideration of an illustrative numerical example.

A.1 SCHEMES FOR THE SOLUTION OF THE IDENTIFICATION EQUATIONS

In order to appreciate fully the salient features of E and the need for special methods of solution, consider the following typical set of equations E, arising out of Theorem 4.1.IA:

E: *Obtain* $\mathbf{u}$ *which satisfies*:

$$\frac{\partial y(x,t;\mathbf{u})}{\partial t} + A[y(x,t;\mathbf{u})] = f(x,t) \quad \text{in } Q, \tag{5.1}$$

$$y(\Sigma) = u_1(\Sigma) \quad \text{on } \Sigma, \tag{5.2}$$

$$y(x,0) = u_2(x) \quad \text{in } \Omega, \tag{5.3}$$

$$-\frac{\partial p(x,t;\mathbf{u})}{\partial t}+A[p(x,t;\mathbf{u})]=y(x,t;\mathbf{u})-z(x,t) \quad \text{in } Q, \tag{5.4}$$

$$p(\Sigma)=0 \quad \text{on } \Sigma, \tag{5.5}$$

$$p(x,T)=0 \quad \text{in } \Omega, \tag{5.6}$$

$$-\frac{\partial p(\Sigma;\mathbf{u})}{\partial \nu_{A^*}}+u_1(\Sigma)-z_1(\Sigma)=0 \quad \text{on } \Sigma, \tag{5.7}$$

$$p(x,0;\mathbf{u})+u_2(x)-z_2(x)=0 \quad \text{in } \Omega. \tag{5.8}$$

Observe that the system of equations (5.1)–(5.6) cannot simultaneously be solved forwards in time, since the initial condition on $p(x,0;\mathbf{u})$ is missing. Nor can the system be solved backwards from $t=T$, since $y(x,T;\mathbf{u})$ is missing. However, if $y(x,T;\mathbf{u})$ could be defined compatibly with (5.1)–(5.8), then the system (5.1)–(5.7) could be solved backward from $t=T$ to recover $y(x,0)=u_2(x)$. On the way, we would obtain (from (5.7)) $u_1(\Sigma)$. We shall present, in Section A.2, a scheme for the compatible determination of $y(x,T;\mathbf{u})$ (with respect to (5.1)–(5.8)). This scheme is based on the fact that there exists a continuous linear transformation between $p(x,T;\mathbf{u})$ and $y(x,T;\mathbf{u})$. We shall see that this transformation derives its representation from the solution to a Riccati-like partial differential equation.

An alternative to the simultaneous solution of (5.1)–(5.8) is available. Recall that the functional to be minimized (as a criterion of identification) is given by

$$J(\mathbf{v})=a(\mathbf{v},\mathbf{v})-2\mathrm{I}(\mathbf{v})+c. \tag{5.9}$$

The $\mathbf{u}\in V$ which minimizes $J(\mathbf{v})$ is characterized by the solution of

$$a(\mathbf{u},\mathbf{v})-\mathrm{I}(\mathbf{v})=0 \tag{5.10}$$

along a trajectory $y(x,t;\mathbf{u})$.

By introducing the adjoint system $p(x,t;\mathbf{u})$, we saw that (5.10) is equivalent to (5.7) and (5.8). We now assert that (5.10) is the formal variational derivative of (5.9) (see Appendix 5.3).

It is this recognition which gives the alternative approach to the simultaneous solution of (5.1)–(5.8).

Define

$$\delta J=a(\mathbf{u},\mathbf{v})-\mathrm{I}(\mathbf{v})=(\mathbf{G}(\mathbf{u}),\mathbf{v}) \tag{5.11}$$

$$\mathbf{G}(\mathbf{u})=\begin{bmatrix} -\dfrac{\partial p(\Sigma;\mathbf{u})}{\partial v_{A^*}}+u_1(\Sigma)-z_1(\Sigma) \\ p(x,0;\mathbf{u})+u_2(x)-z_2(x) \end{bmatrix}. \tag{5.12}$$

We note that

$$\delta J = 0 \Leftrightarrow \mathbf{G} = 0$$

The technique proceeds by relaxing the condition

$$\delta J \sim \mathbf{G} = 0 .$$

This relaxation (or violation) of (5.7) and (5.8) is successively modified by the iterative scheme (at the $k+1^{\text{st}}$ iteration)

$$\mathbf{u}^{k+1} = \mathbf{u}^k + \alpha^k \mathbf{S}^k(\mathbf{G}) . \tag{5.13}$$

In (5.12) and (5.13), $\mathbf{G}$ is the gradient of the functional $J(\mathbf{v})$. $\mathbf{S}^k(\mathbf{G})$ is a conjugate gradient and α^k is the modification of $\mathbf{u}^k$ in the direction $\mathbf{S}^k$. In Section A.3 we shall deal fully with the properties of this iterative scheme. First we consider the Ricatti-like decoupling.

A.1.1 Riccati-like Decoupling

As was indicated in Section A.1, the task at hand is to define, in a way compatible with (5.1)–(5.8), the terminal time condition on the system trajectory, $y(x, T; \mathbf{u})$. We pointed out that ,if such a condition were specified, then $\mathbf{u}$ could be recovered (in principle) from a suitable solution to (5.1)–(5.7).

While $\mathbf{u}$ can be recovered from (5.1)–(5.7), frequently the definition of $y(x, T; \mathbf{u})$ is sufficient. That is, the identification problem would be considered solved on obtaining $y(x, T; \mathbf{u})$. Thus we shall consider our task complete, once having obtained an equation yielding as a result $y(x, T; \mathbf{u})$. In fact, we shall obtain an equation for the evolution of $y(x, T; \mathbf{u})$ with T considered variable. Specifically, consider the characterization of $\mathbf{u}$ for the parabolic system with Dirichlet boundary conditions and error functional PR.A. given by equations (5.1)–(5.8). We give the result as a theorem:

THEOREM 5.1. *Given the system of equations* (5.1)–(5.8) *and the hypothesis of Theorem* 4.1,IA, *and*:

(a) *If* $P(x, \xi, t)$ *satisfies*

$$\frac{\partial P(x,\xi,t)}{\partial t} - A_\xi[P(x,\xi,t)] - A_x[P(x,\xi,t)] - \delta(\xi - x)$$

$$+ \int_{\Gamma_s} \frac{\partial P(x,\Sigma_s)}{\partial \nu_{A_s^*}} \frac{\partial P(\Sigma_s,\xi)}{\partial \nu_{A_s^*}} d\Gamma_s = 0 \quad \text{in } \Omega \times \Omega \times (0,T] , \tag{5.14}$$

$$P(x, \Sigma_s) = P(\Sigma_s, \xi) = 0 , \tag{5.15}$$

$$P(x, \xi, 0) = \delta(x - \xi) . \tag{5.16}$$

(b) *If* $(P.,\cdot,t) \in H^2(\Omega \times \Omega)$ *and* $(\partial P/\partial t)(.,\cdot,t) \in L^2(\Omega \times \Omega)$; $t \in (0, T]$, *then:*

(i) *There exists one and only one* $\hat{y}(\cdot,\cdot) \in L^2(Q)$ such that

$$y(x, T; \mathbf{u}) = \hat{y}(x, T),$$

where $\hat{y}(x,t)$ *is the unique solution of the following linear integral equation of the second kind:*

$$\int_\Omega P(x,\xi,t)\left\{\frac{\partial \hat{y}(\xi,t)}{\partial t} + A[\hat{y}(\xi,t)] - f(\xi,t)\right\}dt = z(x,t) - \hat{y}(\xi,t), \quad (5.17)$$

where (5.17) *is valid in the open set* Q. *The conditions satisfied by* $\hat{y}(x,t)$ *on the closure of* Q *are:*

$$\hat{y}(\Sigma) = z_1(\Sigma), \quad (5.18)$$

$$\hat{y}(x,0) = z_2(x). \quad (5.19)$$

REMARKS ON THEOREM 5.1.

(i) We shall see, in the proof of this theorem, that there exists an affine transformation between $p(\cdot,t)$ and $y(\cdot,t)$, denoted by π:

$$\pi: \quad L^2(\Omega) \to H_1{}^0,$$

$$y(\cdot,t) \to p(\cdot,t).$$

This transformation π has, by the Schwartz kernel theorem, the following representation:

$$p(.,t) = \pi[y(\cdot,t)] = -\int_\Omega P(.,\xi,t)[y(\xi,t;\mathbf{u}) - \hat{y}(\xi,t)]\,d\xi. \quad (5.20)$$

Using (5.1)–(5.6), we obtain the equation which $P(x,\xi,t)$ satisfies, namely (5.13). Now we know that there exists a unique $P(x,\xi,t)$ in virtue of (5.20) and the existence and uniqueness of $p(x,t)$ and $y(x,t;\mathbf{u})$. However, it is nontrivial to establish the class of functions to which $P(\cdot,\cdot,\cdot)$ belongs, since equation (5.14) is nonlinear. In any event, $P(x,\xi,t)$ must satisfy (5.14). The point of this remark is that $\hat{y}(\cdot,\cdot)$ exists in $L^2(Q)$ and is unique. Furthermore, $\hat{y}(x,t)$ must satisfy (5.17), (5.18), and (5.19). Sufficient conditions for the system (5.17), (5.18), and (5.19) to have a unique solution of the appropriate class impose restrictions on the class of functions $P(.,\cdot,\cdot)$. These restrictions are given as hypothesis (*b*) and for which (5.17), (5.18), and (5.19) have a unique solution in $L^2(Q)$.

(ii) The theorem asserts that $y(x,T;\mathbf{u}) = \hat{y}(x,t)$ for all $x \in \Omega$. Equation (5.17) gives the evolution of $\hat{y}(x,T)$ for T considered variable. In the context of lumped parameter systems and in the statistical framework given in Chapter 3, Section A.3, $\hat{y}(x, T)$ is called the filtered estimate. Note that, for T considered variable, the open sets Q and Σ must be given an extended definition. Thus for some $T_E > T$,

$$Q_E = Q \times [T, T_E],$$
$$\Sigma_E = \Sigma \times [T, T_E].$$

The hypothesis on $f(x, t), z(x, t), z_1(\Sigma), z_2(x)$ must be given on the appropriate extended sets:

$$f(.,\cdot) \in L^2(Q_E), \qquad z(\cdot,\cdot) \in L^2(Q_E), \qquad z_1(\cdot) \in L^2(\Sigma_E).$$

In this way we avoid a contradiction. For, as we have seen (Chapter 3, Section A.2.1),

$$y(\cdot, T; u) \notin L^2(\Omega);$$

additionally, it is easy to show that

$$\hat{y}(\cdot, T) \notin L^2(\Omega)$$

for T fixed. However, if T varies, $T < T_E$,

$$\hat{y}(\cdot, T) \in L^2(\Omega), \qquad T \in (0, T_E].$$

PROOF OF THEOREM 5.1. The result of the theorem is that it is possible to find an expression for the missing end condition (in time) for $y(x, t; \mathbf{u})$, $y(x, T; \mathbf{u})$. This expression is $\hat{y}(x, T)$, where $\hat{y}(x, t)$ is the solution of a given integral equation, (5.17). It is possible to find $y(x, T; \mathbf{u})$ because of the existence of a continuous affine mapping from $y(\cdot, t; \mathbf{u}) \to \mathrm{p}(\cdot, t; \mathbf{u})$ for $t \in (0, T]$. Thus, if this mapping can be constructed explicitly, given $p(\cdot, T)$, $y(x, T; \mathbf{u})$ can be recovered. Thus the task of the proof is twofold: first, to show the existence of the map and, second, to give an explicit formula for it. This formulation includes the result that:

$$y(x, T; \mathbf{u}) = \hat{y}(x, T).$$

We assert here that there exists a unique affine continuous map from $y(\cdot, t; \mathbf{u}) \to p(\cdot, t; \mathbf{u})$. That is,

$$\pi[y(\cdot, t; \mathbf{u})] = p(\cdot, t; \mathbf{u}), \qquad t \in (0, T]. \tag{5.21}$$

We demonstrate (5.21) in Appendix 5.1 and in addition show, by construction, that the following lemma holds:

LEMMA 5.1 (L. SCHWARTZ). *Consider the mapping*

$$\pi: \begin{array}{l} L^2(\Omega) \to L^2(\Omega) \\ y(\cdot, t; \mathbf{u}) \to \pi[y(\cdot, t; \mathbf{u})] \end{array} \qquad \text{affine, continuous.}$$

Then there exists a "distribution kernel" $P(x, \xi, t)$ *on* $\Omega \times \Omega$ *such that*:

$$\pi[y(\cdot, t; \mathbf{u})] = -\int_\Omega P(\cdot, \xi, t)\, y(\xi, t)\, d\xi + g(x, t). \tag{5.22}$$

It will be convenient to define $\hat{y}(x, t)$ as the unique solution of the following integral equation (of the first kind):

$$g(x, t) = \int_\Omega P(x, \xi, t)\hat{y}(\xi, t)\, d\xi. \tag{5.23}$$

Thus, by (5.21),

$$p(x, t; \mathbf{u}) = -\int_\Omega P(x, \xi, t)\, [y(\xi, t; \mathbf{u}) - \hat{y}(\xi, t)]\, d\xi. \tag{5.24}$$

We remark that $P(x, \xi, t)$ is unique, since $p(x, t; \mathbf{u})$, $y(x, t; \mathbf{u})$, and $\hat{y}(x, t)$ are. Thus we have established the existence of a unique $P(x, \xi, t)$. In order to obtain a formula for $P(x, \xi, t)$, we shall use (5.24) in (5.4)–(5.8).

As a first result, we observe that (5.24) into (5.5) gives

$$\int_\Omega P(\Sigma_x, \xi)\, [y(\xi, t; \mathbf{u}) - \hat{y}(\xi, t)]\, d\xi = 0,$$

which implies that

$$P(\Sigma_x, \xi) = 0 \tag{5.25}$$

and similarly

$$P(x, \Sigma_\xi) = 0. \tag{5.26}$$

Next, (5.24) into (5.8) gives:

$$-\int_\Omega P(x, \xi, 0)\, [y(\xi, 0; \mathbf{u}) - \hat{y}(\xi, 0)]\, d\xi + u_2(x, 0) - z_2(x, 0) = 0.$$

Thus

$$P(x, \xi, 0) = \delta(\xi - x), \tag{5.27}$$

$$\hat{y}(x, 0) = z_2(x). \tag{5.28}$$

To obtain evolution equations for $P(x, \xi, t)$ and $y(\xi, t)$, put (5.24) into (5.4) and obtain:

$$\int_\Omega \{\frac{\partial P(x, \xi, t)}{\partial t} [y(\xi, t; \mathbf{u}) - \hat{y}(\xi, t)] + P(x, \xi, t)\, [\frac{\partial y(\xi, t; \mathbf{u})}{\partial t} - \frac{\partial \hat{y}(\xi, t)}{\partial t}]$$

$$- A_x[P(x, \xi, t)]\, [y(\xi, t; \mathbf{u}) - \hat{y}(\xi, t)]\}\, d\xi = y(x, t; \mathbf{u}) - z(x, t). \tag{5.29}$$

Equation (5.29) implies (Appendix 5.2) that:

$$\frac{\partial P(x,\xi,t)}{\partial t} - A_\xi[P(x,\xi,t)] - A_x[P(x,\xi,t)] - \delta(\xi - x)$$

$$+ \int_{\Gamma_s} \frac{\partial P(x,\Sigma_s)}{\partial v^*_{A_s}} \frac{\partial P(\Sigma_s,\xi)}{\partial v^*_{As}} d\Gamma_s = 0 \quad \text{in } \Omega\times\Omega\times(0,T], \tag{5.30}$$

$$\int_\Omega P(x,\xi,t)\left\{\frac{\partial \hat{y}(\xi,t)}{\partial t} + A_\xi[\hat{y}(\xi,t)] - f(\xi,t)]\right\}d\xi = z(x,t) - \hat{y}(x,t)$$

$$\text{in } \Omega\times(0,T], \tag{5.31}$$

$$\hat{y}(\Sigma) = z_1(\Sigma)\,. \tag{5.32}$$

Now, for $P(x,\xi,t)$ defined by (5.30), (5.31) has a unique solution $\hat{y}(.,.)\in L^2Q$ provided hypothesis (b) of the theorem holds. Finally, we see that (5.23) substituted into (5.6) implies:

$$-\int_\Omega P(x,\xi,T)\,[y(\xi,T;\mathbf{u}) - \hat{y}(\xi,T)]\,d\xi = 0\,. \tag{5.33}$$

Since $P(x,\xi,t)$ is defined by (5.25), (5.26), (5.27) and (5.30), we must have

$$y(x,T;\mathbf{u}) = \hat{y}(x,T) \quad \text{for all } x\in\Omega\,.$$

We have shown how it is possible to obtain an expression for $y(x,T;\mathbf{u})$ compatible with equations (5.1)–(5.8), and thus it is possible (in principle) to obtain $u_1(\Sigma)$ and $u_2(x)$ by the simultaneous solution of (5.1)–(5.7) "backward" in time, starting with $t = T$. As we remarked, the evolution of $y(x,T;\mathbf{u})$ from T forward (as more measurement information contained in $\varepsilon(x,T)$ and $\varepsilon_1(\Gamma,T)$ becomes available) is often of more interest, especially in the case of the "on-line" control problem. In that situation a refinement $y^*(x,t)$ of the true system state of nature is used to define a control strategy $m(x,t;y^*)$ The important feature of this control is that it is a function of the refinement $y^*(x,t)$ at the "present time," t. If we were to choose $y^*(.,.) = y(\cdot,\cdot;\mathbf{u})$, an inherent time delay, composed of two parts, one due to the data-gathering process and the other due to computation, would be incurred. Thus, at the present time, t, we would have available $y(.,t-\Delta t;\mathbf{u})$, where Δt is the time delay. However, using $y^*(\cdot,t) = \hat{y}(\cdot,t)$, $t\geq T$, there is no data-gathering time delay. We expect that $m(x,t;y)$ would be a better control action than $m(x,t;y(\cdot,t-\Delta;\mathbf{u}))$, providing T is large enough.

There appear to be trade-offs in this context. The relative accuracy of the refinement $y(x,t;\mathbf{u})$ makes its incorporation in the control strategy attractive. While the computation time necessaiy to recover $y(x,t;\mathbf{u})$ via the Ricatti-like transformation just discussed is prohibitive, another technique (to be presented in Section A.1.3) generates $y(x, t; \mathbf{u})$ with relative rapidity. Thus, if $M[y(\cdot, t; \mathbf{u}) - y(\cdot, t - \Delta t)]$ is small ($M[.]$ is some appropriate metric), where Δt is the data-gathering and computation time combined, then it may be advantageous to use the refinement $y(x, t; \mathbf{u})$ in the control scheme rather than $\hat{y}(x, t)$.

Of course, we are interested in $y(x, t; \mathbf{u})$ in its own right, and we presented a justification of the usefullness of the refinement $\hat{y}(x, t)$ in the context of on-line control as a special case. Consequently we return our attention to the problem of solving for $\mathbf{u}$ (and thus $y(x, t; \mathbf{u})$) from the equations characterizing $\mathbf{u}$. In particular, a direct "method" is considered in detail.

A.1.2 Successive Approximation Technique

As we have repeatedly stated, equations (5.1)–(5.8) are the characterization of $\mathbf{u}$ which solves the identification problem. It is worth reviewing the generation of these equations for the purpose of introducing a direct method for their solution.

Recall that the identification problem was phrased as the selection of $\mathbf{v} \varepsilon V$ which extremized a quadratic functional $J(\mathbf{v})$,

$$J(\mathbf{v}): \quad V \to R^1 .$$

By convention, we defined $\mathbf{u}$ implicitly as follows:

$$J(\mathbf{u}) = \inf_{\mathbf{v} \in V} J(\mathbf{v}) = \inf_{\mathbf{v} \in V} \{a(\mathbf{v}, \mathbf{v}) - 2\mathrm{I}(\mathbf{v}) + c\} . \tag{5.34}$$

Under appropriate hypothesis, the unique $\mathbf{u}$ is given by

$$a(\mathbf{u}, \mathbf{v}) - \mathrm{I}(\mathbf{v}) = 0 \quad \text{for all } \mathbf{v} \in V . \tag{5.35}$$

The determination of $\mathbf{u}$ given in (5.34) is nothing more than a mathematical programming problem in the Hilbert space V. There are several iterative schemes for the direct solution of (5.35), each of the following genre:

PA. *Programming Algorithm*

(i) Initial selection of $\mathbf{u}^o \in V$.

(ii) Compute "gradient" of $J(\mathbf{v})$ with respect to $\mathbf{v}$, evaluated at $\mathbf{u}^o$. Denote gradient by $\mathbf{G}$. (The term gradient will be defined presently.)

(iii) If $\mathbf{G}$ is nonzero, then update the initial choice of $\mathbf{u}$ according to:

$$\mathbf{u}^{i+1} = \mathbf{u}^i + \alpha^i \mathbf{S}^i, \qquad i = 0, 1, 2, \ldots ,$$

where $\mathbf{S}^i$ is a meaningful modification of $\mathbf{G}$ and α^i is the magnitude of the step in the direction $\mathbf{S}^i$ on the surface $J(\mathbf{v})$. By meaningful modification of $\mathbf{G}$ we mean that, after n iterations,

$$J(\mathbf{u}^n) \leq J(\overline{\mathbf{u}}^n), \qquad \overline{\mathbf{u}}^o = \mathbf{u}^o ,$$

where $\{\overline{\mathbf{u}}^n\}$ are generated by stepping along $\{\mathbf{G}^n\}$, and $\{\mathbf{u}^n\}$ are generated by stepping along $\{\mathbf{S}^n\}$.

The properties of this computational algorithm are considered in some detail presently. Suffice it to say here that $J(\mathbf{u}^i)$, $i = 0, 1, 2, \ldots$, is a monotone decreasing sequence. Also

$$\lim_{i \to \infty} \mathbf{u}^i \xrightarrow{\text{weakly}} \mathbf{u} \in V .$$

As a meaningful modification to $\mathbf{G}^i$ we chose $\mathbf{S}^i$ as the conjugate gradient [21–23]. Consider, then, the following mathematical programming problem.

A.1.3 The Mathematical Programming Problem Hypothesis

Given a quadratic functional $J(\mathbf{v})$ on the Hilbert space V, $J(\mathbf{v})$: $V \to R^1$,

$$J(\mathbf{v}) = a(\mathbf{v}, \mathbf{v}) - 2\mathrm{I}(\mathbf{v}) + c ,$$

where

$a(\mathbf{v}, \mathbf{v})$ is a continuous bilinear functional of $\mathbf{v} \in V$,
$\mathrm{I}(\mathbf{v})$ is a continuous linear functional of $\mathbf{v} \in V$.

In addition, $a(\mathbf{v}, \mathbf{v})$ is coercive (bounded from below):

$$a(\mathbf{v}, \mathbf{v}) \geq \gamma \|\mathbf{v}\|_V{}^2, \qquad \gamma > 0 .$$

Problem. select $\mathbf{u} \in V$ such that

$$J(\mathbf{u}) = \inf_{v \in V} J(\mathbf{v}) . \tag{5.36}$$

As we have seen Theorem (2.5), there is a unique $\mathbf{u} \in V$ with the property (5.36) and it is characterized by

$$a(\mathbf{u}, \mathbf{v}) - \mathrm{I}(\mathbf{v}) = 0 . \tag{5.37}$$

For a given $\mathbf{u} \in V$, (5.37) is a continuous linear functional of $\mathbf{v}$ and thus has the representation

$$a(\mathbf{u}, \mathbf{v}) - \mathrm{I}(\mathbf{v}) = (\mathbf{G}(\mathbf{u}), \mathbf{v})_V . \tag{5.38}$$

We show in Appendix 5.3 that $G(\mathbf{u})$ is the variational derivative of the functional $J(\mathbf{v})$ evaluated at $\mathbf{u}$. This is what we have previously called the gradient of the functional $J(\cdot)$.

We now specialize the programming algorithm for generating **u** presented earlier as follows:

Algorithm.

(i) Select $\mathbf{u}^o \in V$ (initial guess).

(ii) Evaluate $\mathbf{G}(\mathbf{u}^o)$. If $\mathbf{G}(\mathbf{u}^o) = 0$, then, by (5.37) and (5.38), $\mathbf{u}^o$ is the solution. If $\mathbf{u}^o \neq 0$, then for the $(i+1)$st iteration $(i = 0,1,2,\ldots)$, proceed as follows:

(iii) $\mathbf{u}^{i+1} = \mathbf{u}^i + \alpha^i s^i$,

$$\mathbf{S}^o = -\mathbf{G}(\mathbf{u}^o) ,$$

$$\mathbf{S}^{i+1} = -\mathbf{G}(\mathbf{u}^{i+1}) + \beta^i \mathbf{S}^i , \tag{5.39}$$

$$\beta^i = \frac{(\mathbf{G}(\mathbf{u}^{i+1}), \mathbf{G}(\mathbf{u}^{i+1}))_V}{(\mathbf{G}(\mathbf{u}^i), \mathbf{G}(\mathbf{u}^i))_V} . \tag{5.40}$$

In addition, α^i is chosen so that

$$J(\mathbf{u}^i + \alpha^i \mathbf{S}^i) = \inf_{\gamma^i \in R^1} J(\mathbf{u}^i + \gamma^i \mathbf{S}^i) \tag{5.41}$$

It is possible to obtain an explicit expression for α^i.

$$\alpha^i = -\frac{a(\mathbf{S}^i, \mathbf{u}^i) - \mathrm{I}(\mathbf{s}^i)}{a(\mathbf{S}^i, \mathbf{S}^i)} = \frac{(\mathbf{G}(\mathbf{u}^i), \mathbf{G}(\mathbf{u}^i))_V}{a(\mathbf{S}^i, \mathbf{S}^i)} . \tag{5.42}$$

Before proceeding to show the connection of this technique with equations (5.1)–(5.8), we review some properties of the algorithm in the following theorems.

THEOREM 5.2. *Given the hypothesis of Section* A.1.3, *then, if* $\mathbf{G}(\mathbf{u}^i) \neq 0$, $J(\mathbf{u}^{i+1}) < J(\mathbf{u}^i)$. We prove theorem 5.2 in Appendix 5.5.

COROLLARY 5.2. *The sequence of real numbers* $J(\mathbf{u}^i)$ *is monotone decreasing and hence has a limit in the extended real numbers*:

$$\lim_{i \to \infty} J(\mathbf{u}^i) = J_\infty = \inf_{\mathbf{v} \in V} J(\mathbf{v}) .$$

THEOREM 5.3 (CONVERGENCE). *The sequence* $\{\mathbf{u}^i\}$ *converges weakly to a unique* $\mathbf{u} \in V$ *and the limit* **u** *has the property that*

$$J(\mathbf{u}) = \inf_{\mathbf{v} \in V} J(\mathbf{v}) .$$

That is,

$$\lim_{i \to \infty} \mathbf{u}_i \xrightarrow{\text{weakly}} \mathbf{u} \in V \quad \text{unique},$$

$$J(\mathbf{u}) = J_\infty .$$

PROOF. $\mathbf{u}^i$ is a minimizing sequence and so Theorem 2.3 gives the result.

We have shown that the sequence $\{\mathbf{u}^i\}_{i=1,2,\ldots}$ generated by the algorithm converges (weakly) to a unique $\mathbf{u} \in V$ which realizes the infimum of $J(\mathbf{v})$.

For a discussion of the relative merits of the conjugate gradient direction of search compared with a steepest descent (or gradient technique), see [22].

It is now possible to display the connection between equations (5.1)–(5.8) with the programming technique presented. To be explicit, the functional $J(\mathbf{v})$ associated with (5.1)–(5.8) was given as

$$J(\mathbf{v}) = a(\mathbf{v}, \mathbf{v}) - 2\,\mathrm{I}(\mathbf{v}),$$

with

$$a(\mathbf{v}, \mathbf{v}) = \int_Q (y(x,t;\mathbf{v}) - y(x,t;\mathbf{0}))^2\, dx\, dt + \int_\Sigma (v_1(\Sigma))^2\, d\Sigma + \int_\Omega (v_2(x))^2\, dx,$$

$$\mathrm{I}(\mathbf{v}) = -\Big\{\int_Q (y(x,t;\mathbf{v}) - y(x,t;\mathbf{0}))(y(x,t;\mathbf{0}) - z(x,t))\, dx\, dt - \int_\Sigma u_1(\Sigma)\, z_1(\Sigma)\, d\Sigma - \int_\Omega u_2(x)\, z_2(x)\, dx\Big\}.$$

We saw that the characterization of $\mathbf{u}$ given by

$$a(\mathbf{u}, \mathbf{v}) - \mathrm{I}(\mathbf{v}) = 0 \qquad \text{for all } \mathbf{v} \in V \tag{5.43}$$

could be expressed in terms of an adjoint variable $p(x,t;\mathbf{u})$ (which was introduced for the sole purpose of obtaining such an expression). The expression equivalent to (5.43) was shown to be (equations (4.39), (5.38)):

$$(\mathbf{G}(\mathbf{u}), \mathbf{v})_V = 0 = \int_\Sigma \left[-\frac{\partial p(\Sigma)}{\partial \nu_{A^*}} + u_1(\Sigma) - z_1(\Sigma) \right] v_1(\Sigma)\, d\Sigma + \int_\Omega [p(x,0) + u_2(x) - z_2(x)]\, v_2(x)\, dx .$$

From which we recover

$$\mathbf{G}(\mathbf{u}^i) = \begin{bmatrix} -\dfrac{\partial p(\Sigma;\mathbf{u}^i)}{\partial \nu_{A^*}} + u_1{}^i - z_1{}^i \\ \\ p(x,0;\mathbf{u}^i) + u_2{}^i - z_2{}^i \end{bmatrix}. \tag{5.44, 5.45}$$

$p(x,t;\mathbf{u}^i)$ evolves according to

$$-\frac{\partial p(x,t;\mathbf{u}^i)}{\partial t} + A[p(x,t;\mathbf{u}^i)] = y(x,t;\mathbf{u}^i) - z(x,t), \tag{5.46}$$

$$p(\Sigma) = 0, \tag{5.47}$$

$$p(x,T) = 0, \tag{5.48}$$

and $y(x,t;\mathbf{u}^i)$ evolves according to

$$\frac{\partial y(x,t;\mathbf{u}^i)}{\partial t} + A[y(x,t;\mathbf{u}^i)] = f(x,t), \tag{5.49}$$

$$y(\Sigma) = u_1{}^i, \tag{5.50}$$

$$y(x,0) = u_2{}^i. \tag{5.51}$$

Equations (5.46)–(5.51) are equivalent to (5.1)–(5.8) with $\mathbf{u}$ in the latter replaced by $\mathbf{u}^i$ in the former.

Thus the algorithm would proceed as follows:

(i) Select an initial $\mathbf{u}^0$.

(ii) Compute $\mathbf{G}(\mathbf{u}^0)$. To do so, we first compute $y(.,.;\mathbf{u}^0)$, starting from $u_2{}^0(x,0)$, evolving according to (5.49), under the influence (on the boundary Σ) of $u_1{}^0(\Sigma)$. We remark that this is the numerically stable direction of solution. Having obtained $y(\cdot,\cdot;\mathbf{u}^0)$, it is possible to generate $p(\cdot,\cdot;\mathbf{u}^0)$, starting from $p(x,T)=0$, *evolving backwards in time* according to (5.46). Again we remark that this is the numerically stable direction of solution. Thus $p(\cdot,\cdot;\mathbf{u}^i)$ has been recovered. It is now possible to evaluate $\mathbf{G}(\mathbf{u}^0)$ given by (5.44) and (5.45).

(iii) Using $\mathbf{G}(\mathbf{u}^0)$, compute

$$\mathbf{u}^1 = \mathbf{u}^0 - \alpha^0 \mathbf{G}(\mathbf{u}^0),$$

where

$$\alpha^0 = \frac{(\mathbf{G}(\mathbf{u}^0), \mathbf{G}(\mathbf{u}^0))_V}{a(\mathbf{G}(\mathbf{u}^0), \mathbf{G}(\mathbf{u}^0))}.$$

Explicitly,

$$(\mathbf{G}(\mathbf{u}^0), \mathbf{G}(\mathbf{u}^0))_V = \int_\Sigma \left[-\frac{\partial p(\Sigma;\mathbf{u}^0)}{\partial v_A} + u_1{}^0(\Sigma) - z_1(\Sigma) \right]^2 d\Sigma$$

$$+ \int_\Omega [p(x,0;\mathbf{u}^0) + u_2{}^0(x) - z_2(x)]^2\, dx,$$

$$a(\mathbf{G}(\mathbf{u}^0), \mathbf{G}(\mathbf{u}^0)) = \int_Q [y(x,t;\mathbf{G}(\mathbf{u}^0)) - y(x,t;\mathbf{0})]^2\, dx\, dt$$

$$+ (\mathbf{G}(\mathbf{u}^0), \mathbf{G}(\mathbf{u}^0))_V.$$

The $(i+1)$st iteration proceeds as previously outlined $(i = 1, 2\ldots.)$:

(ii)′ Evaluate $\mathbf{G}(\mathbf{u}^{i+1})$ in a way analogous to that outlined in (ii).
(iii)′ Using $\mathbf{G}(\mathbf{u}^{i+1})$, compute:

$$\mathbf{u}^{i+1} = \mathbf{u}^i + \alpha^i \mathbf{S}^i,$$

$$\alpha^i = \frac{(\mathbf{G}(\mathbf{u}^i), \mathbf{G}(\mathbf{u}^i))_V}{a(\mathbf{S}^i, \mathbf{S}^i)},$$

$$\mathbf{S}^{i+1} = -\mathbf{G}(\mathbf{u}^{i+1}) + \beta^i \mathbf{S}^i,$$

$$\beta^i = \frac{(\mathbf{G}(\mathbf{u}^{i+1}), \mathbf{G}(\mathbf{u}^{i+1}))_V}{(\mathbf{G}(\mathbf{u}^i), \mathbf{G}(\mathbf{u}^i))_V}.$$

The explicit representation formulas for the quantities α^i, β^i, etc., are now obvious.

The iteration continues until $(\mathbf{G}(\mathbf{u}^i), \mathbf{G}(\mathbf{u}^i))_V$ is acceptably close to zero.

In Sections A.1.1 and A.1.3 we have outlined two distinct methods for obtaining the solution to the simultaneous equations characterizing $\mathbf{u} \in V$ for a specific identification problem. That problem was, as we saw, concerned with obtaining $\mathbf{u} \in V$ which minimized the quadratic error functional $J(\mathbf{v})$ along a system trajectory $y(\cdot,\cdot,\mathbf{v})$, where $y(x,t;\mathbf{v})$ satisfied a parabolic evolution equation with Dirichlet boundary conditions. The other (non-trivial) identification problems treated in Chapter 4 can be solved by the methods of Section A.1.1 and A.1.3 of this chapter, the results being different in detail only. However, the following remarks concerning the direct method of Section A.3 are in order.

(i) The response of the parabolic system S, denoted by $y(.,.;\mathbf{v})$, is stable, in the sense that the homogeneous solution starting from a non zero initial condition $u_2(x,0)$ at $t = 0$ approaches zero as time increases.

(ii) The response of the adjoint parabolic system S^*, denoted by $p(\cdot, \cdot; \mathbf{v})$, *is unstable* in the sense considered in (i). However, the response $p(\cdot, \cdot; \mathbf{v})$ is stable in reverse time, the homogeneous solution starting from a nonzero terminal condition $p(x, T)$ at $t = T$ approaches zero *as time decreases.*

(iii) For the second-order hyperbolic systems the response of the system S or its adjoint S^* is marginally stable in either time direction. (For a particular $x \in \Omega$, $y(x, \cdot; \mathbf{v})$, and $p(x, \cdot; \mathbf{v})$ are a sum of harmonic functions in time).

With these remarks we observe that the numerical computation of the algorithm posed in Section A.1.3 is always stable, since S is solved forwards in time and S^* is solved backwards in time.

The numerical solution of the integro-partial and partial differential equations of Sections A.1.1 and A.1.3, respectively, is accomplished by the approximation technique of Galerkin, the application of which is now considered.

A.2 THE GALERKIN TECHNIQUE

We address ourselves at first to the solution of equations (5.14)–(5.19), repeated here for convenience:

$$\frac{\partial P(x,\xi,t)}{\partial t} - A_\xi[P(x,\xi,t)] - A_x[P(x,\xi,t)] - \delta(\xi - x)$$

$$+ \int_{\Gamma_s} \frac{\partial P(x,\Sigma_s)}{\partial \nu_{A_s^*}} \frac{\partial P(\Sigma_s,\xi)}{\partial \nu_{A_s^*}}\, d\Gamma_s = 0 \quad \text{in } \Omega \times \Omega \times (0,T], \tag{5.14}$$

$$P(x,\Sigma_s) = P(\Sigma_s,\xi) = 0 \quad \text{on } \Sigma_s \times \Omega, \tag{5.15}$$

$$P(x,\xi,0) = \delta(x - \xi) \quad \text{in } \Omega \times \Omega. \tag{5.16}$$

$$\int_\Omega P(x,\xi,t)\left\{\frac{\partial \hat{y}(\xi,t)}{\partial t} + A_\xi[\hat{y}(\xi,t)] - f(\xi,t)\right\} d\xi = z(x,t) - \hat{y}(x,t)$$

$$\text{in } \Omega \times (0,T], \tag{5.17}$$

$$\hat{y}(\Sigma) = z_1(\Sigma) \quad \text{on } \Sigma, \tag{5.18}$$

$$\hat{y}(x,0) = z_2(x) \quad \text{in } \Omega. \tag{5.19}$$

By hypothesis (b) of Theorem 5.1, $P(\cdot, \cdot\ t) \in H^2(\Omega \times \Omega)$, $P(\cdot, \cdot\ t) \in L^2(\Omega \times \Omega)$.

Thus there exists an approximation to $P(x, \xi, t)$, denoted by $P_m(x, \xi, t)$ given by

$$P_m(x, \xi, t) = \sum_{i,\, j=1}^{m} P_{ij}(t) w_i(x) w_j(\xi), \tag{5.52}$$

where $\{w_i(.)w_j(.)\}_{i,j=1,2,\ldots}$ is an orthonormalized basis in $L^2(\Omega \times \Omega)$, $P_m(\cdot, \cdot, t)$ has the property that, for each $t \in (0, T]$,

$$\lim_{m \to \infty} P_m(.,.,t) \to P(\cdot, \cdot, t) \quad \text{in } L^2(\Omega \times \Omega).$$

In particular, we choose the orthonormal basis $\{w_i(\cdot)w_j(\cdot)\}_{i,j=1,2\ldots}$, afforded by the normalized eigenfunctions which are generated as solutions to

$$A_x[w_i(x)] + A_\xi[w_j(\xi)] - (\lambda_i + \lambda_j) w_i(x) w_j(\xi) = 0,$$
$$w_i(\Gamma_x) = 0,$$
$$w_j(\Gamma_\xi) = 0, \qquad i,j = 1, 2, \ldots .$$

Define

$$\Psi(x, \xi, t) = w_k(x) w_l(\xi) g(t), \qquad k, l \text{ fixed but arbitrary},$$
$$g(t) \in C^1, \qquad g(T) = 0.$$

An equation defining the coefficients $P_{ij}(t)$ of (5.52) can be obtained via the Galerkin approximation scheme, We proceed as follows:

Multiply (5.14) by $\Psi(x, \xi, t)$ and integrate over $(\Omega \times \Omega) \times (0, T]$. There results:

$$\int_0^T \Bigg\{ -\frac{dg(t)}{dt} P_{kl}(t) - (\lambda_k + \lambda_l) P_{kl}(t) g(t) - \Delta_{kl} g(t)$$
$$+ \int_{\Gamma_s} \left[\sum_{j=1}^{m} P_{kj}(t) \frac{\partial w_j(\Gamma_s)}{\partial \nu_{A_s^*}} \right] \left[\sum_{i=1}^{m} P_{il}(t) \frac{\partial w_i(\Gamma_s)}{\partial \nu_{A_s^*}} \right] g(t)\, d\Gamma_s \Bigg\} dt,$$
$$- g(0) \Delta_{kl} = 0, \tag{5.53}$$

where

$$\Delta_{kl} = \begin{cases} 1 & \text{if } k = l, \\ 0 & \text{otherwise}. \end{cases}$$

Integrate the first term of (5.4) by parts and collect terms in $g(t)$. Owing to the arbitrariness of $g(t)$ and the fact that $g(t) \in C^1$, we obtain an equation for $P_{kl}(t)$, $(k, l = 1, 2, \ldots, m)$:

$$\frac{dP_{kl}}{dt} - (\lambda_k + \lambda_l)P_{kl}(t) - \Delta_{kl}$$

$$+ \int_{\Gamma_s} \left[\sum_{j=1}^{m} P_{kj}(t) \frac{\partial w_j(\Gamma_s)}{\partial \nu_{A_s*}} \right] \left[\sum_{i=1}^{m} P_{il}(t) \frac{\partial w_i(\Gamma_s)}{\partial \nu_{A_s*}} \right] d\Gamma_s = 0, \tag{5.54}$$

$$P_{kl}^{(0)} = \Delta_{k1} . \tag{5.55}$$

Equation (5.54) can be solved by any of the many numerical integration schemes available, starting with the initial condition (5.55). Thus, using (5.52) we can obtain $P_m(., ., .), x, \xi, t \in \Omega \times \Omega[0, T]$.

Equation (5.17) is a linear integral equation of the second kind. Having obtained the solution $P_m(x, \xi, t)$, (5.17) can be solved by the sucessive substitution or successive approximation methods [15]. Rather than proceed in that manner, however (because we seek an equation for *the evolution* of $\hat{y}_i(t)$, not $\hat{y}_i(t)$), consider the following approach: Multiply (5.17) by $g(t)$ and integrate over $(0, T]$ to obtain:

$$\int_Q \left\{ -\frac{\partial}{\partial t}[P(x, \xi, t)g(t)] + g(t)A_\xi[P(x, \xi, t)] \right\} \hat{y}(\xi, t)\, d\xi\, dt$$

$$- \int_Q f(\xi, t) P(x, \xi, t) g(t)\, d\xi\, dt - \int_\Omega P(x, \xi, 0)\, \hat{y}(\xi, 0) g(0)\, d\xi$$

$$+ \int_{\Sigma_s} \left[y(\Sigma_s) \frac{\partial P(x, \Sigma_s)}{\partial \nu_{A_s*}} g(t)\, d\Sigma_s = \int_0^T [z(x, t) - \hat{y}(x, t)]\, g(t)\, dt . \tag{5.56}$$

We seek an approximate solution to (5.56) denoted by $\hat{y}_m(x, t)$, where, as before,

$$\hat{y}_m(x, t) = \sum_{i=1}^{m} \hat{y}_i(t) w_i(x), \qquad \lim_{m \to \infty} \hat{y}_m(\cdot, \cdot) \to \hat{y}(\cdot, \cdot) \text{ in } L^2(Q) .$$

Substituting the expression for $y_m(x, t)$ into (5.56) along with the definition of $P_m(x, \xi, t)$ given by (5.52) yields:

$$\int_Q \left\{ - \sum_{k,l=1}^{m} \frac{\partial}{\partial t} [P_{kl}(t) g(t)] w_k(x) w_l(\xi) \sum_{i=1}^{m} \hat{y}_i(t) w_i(\xi) \right.$$

$$+ g(t) \sum_{i=1}^{m} y_i(t) w_i(\xi) \sum_{k,l=1}^{m} P_{kl}(t) w_k(x) A[w_l(\xi)]$$

$$\left. - f(\xi, t) g(t) \sum_{k,l=1}^{m} P_{kl}(t) w_k(x) w_l(\xi) \right\} d\xi \, dt - z_2(x) g(0)$$

$$+ \int_{\Sigma_s} \left[z_1(\Sigma_s) \sum_{k,l=1}^{m} P_{kl}(t) w_k(x) \frac{\partial w_l(\Gamma_s)}{\partial \nu_{A_s^*}} g(t) \, d\Sigma_s \right.$$

$$= \int_0^T [z(x,t) - \hat{y}(x,t)] \, g(t) \, dt. \tag{5.57}$$

Using the orthonormal properties of the set $\{w_i(\cdot)\}_{i=1,2,\ldots}$ and the intentioned arbitrariness of $g(\cdot)$ in $C^1[0, T)$, then (5.57) is equivalent to:

$$\sum_{k,l=1}^{m} P_{kl}(t) w_k(x) \left[\frac{d\hat{y}_l}{dt} + \lambda_l \hat{y}_l(t) - f_l(t) + z_{1l}(t) \right] = \hat{y}(x,t) - z(x,t), \tag{5.58}$$

$$- z_2(x) + \sum_{k,l=1}^{m} P_{kl}(0) \hat{y}_l(0) w_k(x) = 0, \tag{5.59}$$

where

$$z_{1l}(t) = \int_{\Gamma_s} z_1(\Sigma_s) \frac{\partial w_l}{\partial \nu_{A_s^*}} \, d\Gamma_s,$$

$$f_l(t) = \int_{\Omega} f(x,t) w_l(x) \, dx.$$

Before proceeding with (5.58) and (5.59), define the following matrix and vector quantities:

$$\mathbf{P} = \{P_{ij}(t)\}_{i=j=1,2,\ldots,m},$$

$$\mathbf{A} = \begin{bmatrix} \lambda_1 & 0 & \ldots & 0 & 0 \\ 0 & \lambda_2 & \ldots & 0 & 0 \\ \vdots & \vdots & \ddots & \vdots & \\ 0 & 0 & \ldots & \lambda_{m-1} & 0 \\ 0 & 0 & \ldots & 0 & \lambda_m \end{bmatrix},$$

$$\mathbf{f} = [f_1(t) \;\; f_2(t) \;\; \ldots \;\; f_m(t)]^T,$$

$$\hat{\mathbf{y}} = [\hat{y}_1(t) \quad \hat{y}_2(t) \quad \ldots \quad \hat{y}_m(t)]^T$$
$$\mathbf{z}_1 = [z_{11}(t) \quad z_{12}(t) \quad \ldots \quad z_{1m}(t)]^T ,$$
$$\mathbf{z}_2 = [z_{21} \quad z_{22} \quad \ldots \quad z_{2m}]^T ,$$

$$\mathbf{z} = [z_1(t) \quad z_2(t) \quad \ldots \quad z_m(t)], \qquad z_i = \int_\Omega z(x,t)w_i(x)\,dx ,$$

$$\mathbf{W}\mathbf{W}^T = \int_{\Gamma_s} \begin{bmatrix} \frac{\partial w_1(\Gamma_s)}{\partial v_{A^*}} \frac{\partial w_1(\Gamma_s)}{\partial v_{A^*}} & \cdots & \frac{\partial w_1(\Gamma_s)}{\partial v_{A^*}} \frac{\partial w_m(\Gamma_s)}{\partial v_{A^*}} \\ \vdots & & \vdots \\ \frac{\partial w_m(\Gamma_s)}{\partial v_{A^*}} \frac{\partial w_1(\Gamma_s)}{\partial v_{A^*}} & \cdots & \frac{\partial w_m(\Gamma_s)}{\partial v_{A^*}} \frac{\partial w_m(\Gamma_s)}{\partial v_{A^*}} \end{bmatrix} d\Gamma_s .$$

Then, if (5.58) and (5.59) are each multiplied by $w_i(x)$ and the results are integrated over Ω, the following equations, given in matrix-vector form, result

$$\mathbf{P}(t)\left[\frac{d\hat{\mathbf{y}}(t)}{dt} + \mathbf{A}\hat{\mathbf{y}}(t) - \mathbf{f}(t) + \mathbf{z}_1(t)\right] = \mathbf{z}(t) - \hat{\mathbf{y}}(t), \tag{5.60}$$

$$\hat{\mathbf{y}}(0) = \mathbf{z}_2 . \tag{5.61}$$

For completeness, we give the matrix representation for $\{P_{ij}(t)\}$ satisfying (5.54) and (5.55):

$$\frac{d\mathbf{P}(t)}{dt} - \mathbf{A}\mathbf{P}(t) - \mathbf{P}(t)\mathbf{A} - \mathbf{I} + \mathbf{P}(t)\mathbf{W}\mathbf{W}^T\mathbf{P}(t) = 0, \tag{5.62}$$

$$\mathbf{P}(0) = \mathbf{I} . \tag{5.63}$$

In order to solve (5.60) for $\hat{\mathbf{y}}(t)$, we need an equation for $\mathbf{P}^{-1}(t)$. Consider the following identity:

$$\frac{d(\mathbf{P}^{-1}(t))}{dt} = -\mathbf{P}^{-1}(t)\frac{d\mathbf{P}(t)}{dt}\mathbf{P}^{-1}(t).$$

Formally, then, pre- and postmultiplication of (5.62) by $\mathbf{P}^{-1}(t)$ gives an equation for the evolution of $\mathbf{P}^{-1}(t)$:

$$\frac{d(\mathbf{P}^{-1}(t))}{dt} + \mathbf{P}^{-1}(t)\mathbf{A} + \mathbf{A}\mathbf{P}^{-1}(t) + \mathbf{P}^{-1}(t)\mathbf{I}\mathbf{P}^{-1}(t) - \mathbf{W}\mathbf{W}^T = 0, \tag{5.64}$$

$$\mathbf{P}^{-1}(0) = \mathbf{I} . \tag{5.65}$$

The existence of an unique solution to (5.64) and (5.65) is given by the following theorem due to Kalman [27].

THEOREM 5.4 (KALMAN). *If* $\mathbf{P}^{-1}(0)$ *is positive definite, then* $\mathbf{P}^{-1}(t)$ *exists for all* $t > 0$ *and is the unique solution of* (5.64) *having the initial value* $\mathbf{P}^{-1}(0)$ *given by* (5.65).

Consequently, since $\mathbf{P}^{-1}(0) = I > 0$, a solution to (5.64) exists and is unique. Thus we can write (5.60) as follows:

$$\frac{d\hat{\mathbf{y}}(t)}{dt} + \mathbf{A}\hat{\mathbf{y}}(t) - \mathbf{f}(t) + \mathbf{z}_1(t) = \mathbf{P}^{-1}(t)[\mathbf{z}(t) - \hat{\mathbf{y}}(t)], \tag{5.66}$$

$$\hat{\mathbf{y}}(0) = \mathbf{z}_2 . \tag{5.67}$$

Rewriting (5.64) and (5.65) here for convenience, $\mathbf{P}^{-1}(t)$ is given by:

$$\frac{d(\mathbf{P}^{-1}(t))}{dt} + \mathbf{P}^{-1}(t)\mathbf{A} + \mathbf{A}\mathbf{P}^{-1}(t) + \mathbf{P}^{-1}(t)\mathbf{I}\mathbf{P}^{-1}(t) - \mathbf{W}\mathbf{W}^T = 0, \tag{5.68}$$

$$\mathbf{P}(0)^{-1} = \mathbf{I} . \tag{5.69}$$

$\hat{y}_m(x, t)$ is obtained by solving (5.66) and (5.68) by a numerical integration scheme, thereby obtaining $\hat{y}_i(t)$, and

$$\hat{y}_m(x, t) = \sum_{i=1}^{m} \hat{y}_i(t) w_i(x) .$$

The numerical technique for solving the equations arising in Section A.2 has been demonstrated. In summary, the partial differential and integral equations were solved by approximation of the solutions in terms of eigenfunction expansions. These expansions were carried out according to the technique of Galerkin (Chapter 2, Section A.3). The expansion led to the consideration of a system of ordinary differential equations, given by (5.66), (5.67), (5.68) and (5.69). We established as a subsidiary result that the system (5.68) and (5.69) had a unique solution, hence was equivalent to the system (5.62) and (5.63).

The equations arising out of the "direct method" of Section A.I.3 can be solved using the Galerkin approximation technique. The method is straightforward and the results can be outlined.

(a) As before, we have the existence of $y_m(x, t; \mathbf{u}^k)$ and $p_m(x, t; \mathbf{u}^k)$, where

$$y_m(x, t; \mathbf{u}^k) = \sum_{i=1}^{m} y_i(t; \mathbf{u}^k) w_i(x),$$

$$p_m(x,t;\mathbf{u}^k) = \sum_{i=1}^{m} p_i(t;\mathbf{u}^k)w_i(x),$$

$$\lim_{m\to\infty} y_m(x,t;\mathbf{u}^k) \to y(x,t;\mathbf{u}^k)$$

$$\lim_{m\to\infty} p_m(x,t;\mathbf{u}^k) \to p(x,t;\mathbf{u}^k).$$

(b) $y_i(t;\mathbf{u}^k)$ and $p_i(t;\mathbf{u}^k)$ are defined, according to Galerkin, by:

$$\frac{dy_i(t;\mathbf{u}^k)}{dt} + \lambda_i y_i(t;\mathbf{u}^k) = f_i(t) - u_{1i}^k(t),$$

$$y_i(0) = u_{2i}^k,$$

$$-\frac{dp_i(t;\mathbf{u}^k)}{dt} + \lambda_i p_i(t;\mathbf{u}^k) = y_i(t;\mathbf{u}^k) - z_i(t),$$

$$p_i(T) = 0$$

with

$$f_i(t) = \int_\Omega f(x,t)w_i(x)\,dx,$$

$$z_i(t) = \int_\Omega z(x,t)w_i(x)\,dx,$$

$$u_{1i}^k(t) = \int_\Gamma u_1^k(\Sigma)\frac{\partial w_i(\Gamma)}{\partial \nu_{A^*}}\,d\Gamma,$$

$$u_{2i}^k = \int_\Omega u_2^k(x)w_i(x)\,dx.$$

(c) The gradient vector $\mathbf{G}(\mathbf{u}^k)$ is given by:

$$\mathbf{G}(\mathbf{u}^k) = \begin{bmatrix} -\sum_{i=1}^{m} p_i(t)\dfrac{\partial w_i(\Gamma)}{\partial \nu_{A^*}} + u_2^k(\Sigma) - z_1(\Sigma) \\ \\ \sum_{i=1}^{m} p_i(0)w_i(x) + u_2^k(x) - z_2(x) \end{bmatrix}.$$

The algorithm outlined in Section A.1.3 is carried out with $y(x,t)$ and $p(x,t)$ replaced by $y_m(x,t)$ and $p_m(x,t)$. The specific technique is best illustrated by the example considered in Section A.3. In that section, two illustrative examples are considered, each of which are solved by the methods described in section A.1.1 and A.1.3.

The problem which was posed in section A.1 was the identification problem with Dirichlet boundary conditions and quadratic functional PR.A defined in Chapter 4. One of the illustrative examples of Section A.3 is concerned with the same system and boundary conditions, but which the quadratic fuctional of PR.B, Chapter 4. That is, the case of pointwise output measurements in the spatial domain. Thus, for completeness, we give a theorem similar in content to Theorem 5.1. The system of equations characterizing the unique $\mathbf{u} \in V$ which solves PR.IB of Chapter 4 is given by Theorem 4.1,IB, and is repeated here:

$$\frac{\partial y(x,t;\mathbf{u})}{\partial t} + A[y(x,t;\mathbf{u})] = f(x,t), \tag{5.70}$$

$$y(\Sigma) = u_1(\Sigma), \tag{5.71}$$

$$y(x,0) = u_2(x). \tag{5.72}$$

$$\frac{\partial p(x,t;\mathbf{u})}{\partial t} + A[p(x,t;\mathbf{u})] = \sum_{i=1}^{\nu} [y(x,t;\mathbf{u}) - z(x,t;\mathbf{u})]\,\delta(x-x^i) \quad \text{in } Q, \tag{5.73}$$

$$p(\Sigma;\mathbf{u}) = 0, \tag{5.74}$$

$$p(x,T) = 0, \tag{5.75}$$

$$-\frac{\partial p(\Sigma;\mathbf{u})}{\partial \nu_{A^*}} + u_1(\Sigma) - z_1(\Sigma) = 0, \quad \text{on } \Sigma, \tag{5.76}$$

$$p(x,0;\mathbf{u}) + u_2(x) - z_2(x) = 0, \quad \text{in } \Omega. \tag{5.77}$$

THEOREM 5.5. *Given the system of equations* (5.70)–(5.77) and the *hypothesis of Theorem* 4.1,IB, *and*:

(a) *If* $P(x\,\xi, t,)$ *satisfies*

$$\frac{\partial P(x,\xi,t)}{\partial t} - A_\xi[P(x,\xi,t)] - A_x[P(x,\xi,t)] - \sum_{i=1}^{\nu} \delta(\xi - x)\delta(\xi - x^i) + \int_{\Gamma_s} \frac{\partial P(x,\Sigma_s)}{\partial \nu_{A_s^*}} \frac{\partial P(\Sigma_s,\xi)\,d\Gamma}{\partial \nu_{A_s^*}} = 0, \tag{5.78}$$

$$P(x,\Sigma_s) = P(\Sigma_s,\xi) = 0, \tag{5.79}$$

$$P(x,\xi,0) = \delta(\xi - x). \tag{5.80}$$

(b) *If* $P(.,.;t) \in H^2(\Omega \times \Omega)$ *and* $\partial p/\partial t)(\cdot,\cdot;t) \in L^2(\Omega \times \Omega)$ *for each* $t(0,T]$, *then*

(i) *There exists one and only one* $y(\cdot,\cdot) \in L^2(Q)$ *such that*

$$y(x,T;\mathbf{u}) = \hat{y}(x,T),$$

where $\hat{y}(x,t)$ *is the unique solution of the following linear integral equation of the second kind*:

$$\int_\Omega P(x,\xi,t)\left[\frac{\partial \hat{y}(\xi,t)}{\partial t} + A_\xi[\hat{y}(\xi,t)] - f(\xi,t)\right] d\xi$$

$$= \sum_{i=1}^{\nu} [\hat{y}(x,t) - z(x,t)]\,\delta(x - x^i), \tag{5.81}$$

where (5.81) *is valid in the partially open set* Q. *The conditions satisfied by* $\hat{y}(\xi,t)$ *on the closure of* Q *are*:

$$\hat{y}(\Sigma) = z_1(\Sigma), \tag{5.82}$$

$$y(x,0) = z_2(x). \tag{5.83}$$

(ii) *The equations of* (i) *can be approximated in the following way*:

$$\hat{y}_m(x,t) = \sum_{i=1}^{m} \hat{y}_i(t) w_i(x),$$

$$P_m(x,\xi,t) = \sum_{i=j=1}^{m} P_{ij}(t) w_i(x) w_j(x),$$

which have the property

$$\left.\begin{aligned} &\lim_{m\to\infty} \hat{y}_m(.,t) \to \hat{y}(.,t) && \text{almost everywhere in } L^2(\Omega), \\ &\lim_{m\to\infty} P_m(\cdot,\cdot,t) \to P(\cdot,\cdot,t) && \text{almost everywhere in } L^2(\Omega\times\Omega), \end{aligned}\right\} \text{for each } t \in (0,T],$$

$\{w_i(\cdot)\}_{i=1,2,\ldots}$ *are an arbitrary orthonormal basis in* $L^2(\Omega)$.

The coefficients $\hat{y}_i(t)$ and $P_{ij}(t)$ satisfy the following equations:

$$\frac{d\hat{\mathbf{y}}(t)}{dt} + \mathbf{A}\hat{\mathbf{y}}(t) - \mathbf{f}(t) + \mathbf{z}_1(t) = \mathbf{P}^{-1}(t)\mathbf{Q}_\nu(\mathbf{z}(t) - \hat{\mathbf{y}}(t)), \tag{5.84}$$

$$\hat{\mathbf{y}}(0) = \mathbf{z}_2, \tag{5.85}$$

$$\frac{d(\mathbf{P}^{-1}(t))}{dt} + \mathbf{P}^{-1}(t)\mathbf{A} + \mathbf{A}\mathbf{P}^{-1}(t) + \mathbf{P}^{-1}(t)\mathbf{Q}_\nu\mathbf{P}^{-1}(t) - \mathbf{W}\mathbf{W}^T = 0, \tag{5.86}$$

$$\mathbf{P}^{-1}(0) = \mathbf{I}, \tag{5.87}$$

where the matrices and vectors appearing in (5.84)–(5.87) are as previously defined with the exception of Q_v, defined below:

$$\mathbf{Q}_v = \sum_{i=1}^{v} \mathbf{Q}^i, \qquad \mathbf{Q}^i = \{q_{kl}^i\}_{k,l=1,2,\ldots,m},$$

$$q_{kl}^i = w_k(x^i)\,w_l(x^i).$$

Remarks on Theorem 5.5. (i) The proof of this theorem is the same as for Theorem 5.1 and is not repeated.

(ii) The difference in the equations for $\hat{y}(x,t)$ (the case of continuous interior measurements and discrete interior measurements), is encapsulated by a comparison of (5.66) and (5.67) with (5.84) and (5.85), respectively. In the latter case the feedback error term on the R. H. S. of (5.84) reflects weighted sum errors arising at the discrete measurement locations x^i $(i = 1, 2, \ldots, v)$.

The numerical solution of two illustrative examples is considered next. These examples serve to consolidate the theoretical results.

A.3 NUMERICAL EXAMPLES

Two numerical examples serve as vehicles for the display of some salient features of the solution technique presented in Sections A.1 and A.2. In each case the system is of parabolic type with Dirichlet boundary conditions. The point of departure between the examples is in the definition of the output measurement process, and in consequence a direct comparison between the resulting refined estimates is afforded. In the first case, output measurements are taken for each $t \in (0, T]$, over the entire spatial profile Ω. In the second case, output measurements are taken for each $t \in (0, T]$ at selected points along the spatial profile denoted by x^i, $x^i \in \Omega$ $(i = 1, 2, \ldots, v)$.

Consider then the following two examples in turn.

Example 1

System S_D.

$$\frac{\partial y(x,t;\mathbf{u})}{\partial t} - \frac{\partial^2 y(x,t)}{\partial x^2} = 212.0, \qquad x, t \in (0,1) \times (0,1]. \tag{5.88}$$

$$y(0,t) = u_1(0,t), \qquad t \in (0,T], \tag{5.89}$$

$$y(1,t) = u_1(1,t), \qquad t \in (0,T], \tag{5.90}$$

$$y(x,0) = u_2(x), \qquad x \in (0,1). \tag{5.91}$$

Input Measurements I.

$$z_1(0,t) = u_1^*(0,t)+k_1N_1(t), \qquad t\in(0,1]\,, \tag{5.92}$$

$$z_1(1,t) = u_1^*(1,t)+k_2N_2(t), \qquad t\in(0,1]\,, \tag{5.93}$$

$$z_2(x) \;\; = u_2^*(x) \;\; +k_3, \qquad x\in(0,1)\,, \tag{5.94}$$

where $\mathbf{u}^*$ is the true state of nature, and is defined by:

$$u_1^*(0,t) = 70+10\sin 2\Pi t\,, \tag{5.95}$$

$$u_1^*(1,t) = 54.5\,, \tag{5.96}$$

$$u_2^*(x) \;\; = 70e^{-0.25x}\,, \tag{5.97}$$

and
$N_1(t)$, $N_2(t)$ are purely random functions of time, each independent of the other with amplitude ± 1.0. k_1 and k_2 are variables chosen to affect the signal to noise ratio. k_3 is an arbitrary bias on the initial condition, also chosen to affect the signal to noise ratio.

Evidently, from (5.92)–(5.97) and the fact that $k_1N_1(t)$ and $k_2N_2(t)$ are amplitude limited:

$$z_1(\cdot)\in L^2(\Sigma)\,,$$

$$z_2(\cdot)\in L^2(\Omega)\,.$$

Output Measurements O.

$$z(x,t) = y(x,t;\mathbf{u}^*)+k_0N_0(t), \qquad x,t\in(0,1)\times(0,1]\,, \tag{5.98}$$

where $y(x,t;\mathbf{u}^*)$ is the response of the system S_D at the point $x,t\in(0,1)\times(0,1]$ which evolves according to (5.88) with initial condition $u_2^*(x)$ and boundary conditions $u_1^*(0,t)$ and $u_1^*(1,t)$. $N_0(t)$ is a purely random function of time with amplitude ± 1.0. The set $\{N_0(t), N_1(t), N_2(t)\}$ are pairwise independent for all $t\in(0,1]$. k_0 is a variable which alters the signal-to-noise ratio.

The problem is to choose $\mathbf{u}\in V$ which extremizes an error functional $J(\mathbf{v})$, $\mathbf{v}\in V$.

Define

$$V = L^2(\Sigma)\in L^2(\Omega)\,,$$

$$J(\mathbf{v}) = \int_0^1\int_0^1 [y(x,t;\mathbf{v})-z(x,t)]^2\,dx\,dt$$

$$+\int_0^1 [v_1(0,t)-z_1(0,t)]^2\,dt+\int_0^1 [v_1(1,t)-z_1(1,t)]^2\,dt$$

$$+\int_0^1 [v_2(x)-z_2(x)]^2\,dx\,. \tag{5.99}$$

As we have seen (Theorem 4.1,IA), the unique $\mathbf{u} \in V$ which extremizes $J(\mathbf{v})$ is characterized by:

$$\frac{\partial y(x,t;\mathbf{u})}{\partial t} - \frac{\partial^2 y}{\partial x^2} = 212.0, \qquad x,t \in (0,1) \times (0,1], \tag{5.100}$$

$$y(0,t) = u_1(0,t), \qquad t \in (0,1], \tag{5.101}$$

$$y(1,t) = u_1(1,t), \qquad t \in (0,1], \tag{5.102}$$

$$y(x,0) = u_2(x) \qquad x \in (0,1). \tag{5.103}$$

$$\frac{\partial p(x,t;\mathbf{u})}{\partial t} - \frac{\partial^2 p(x,t;\mathbf{u})}{\partial x^2} = y(x,t;\mathbf{u}) - z(x,t), \tag{5.104}$$

$$p(0,t) = 0, \tag{5.105}$$

$$p(1,t) = 0, \tag{5.106}$$

$$p(x,1) = 0. \tag{5.107}$$

$$\frac{\partial p(0,t;\mathbf{u})}{\partial x} + u_1(0,t) - z_1(0,t) = 0, \quad t \in (0,1], \tag{5.108}$$

$$-\frac{\partial p(1,t;\mathbf{u})}{\partial x} + u_1(1,t) - z_1(1,t) = 0, \quad t \in (0,1], \tag{5.109}$$

$$p(x,0;\mathbf{u}) + u_2(x) - z_2(x) = 0, \quad x \in (0,1). \tag{5.110}$$

The system state $y(x,t;\mathbf{u})$ and its adjoint $p(x,t;\mathbf{u})$ were approximated, in accordance with the developments of Section A.2, by:

$$y_m(x,t;\mathbf{u}) = \sum_{i=1}^{4} y_i(t;\mathbf{u}) w_i(x), \tag{5.111}$$

$$p_m(x,t;\mathbf{u}) = \sum_{i=1}^{4} p_i(t;\mathbf{u}) w_i(x), \tag{5.112}$$

$$\hat{y}_m(x,t) = \sum_{i=1}^{8} \hat{y}_i(yt) w_i(x), \tag{5.113}$$

$$w_i(x) = \sqrt{2}\,\sin(\sqrt{\lambda_i})x, \qquad \lambda_i = (i\Pi)^2. \tag{5.114}$$

Remark.

$$\{\sqrt{2}\sin(i\Pi)(\cdot)\}_{i=1,2,\ldots} \text{ are complete in } L^2(\Omega),$$

$$\int_0^1 \sqrt{2}\sin(i\Pi)x\sqrt{2}\sin(j\Pi)x\,dx = \begin{matrix} 1 & \text{if } i=j, \\ 0 & \text{otherwise}. \end{matrix}$$

Thus, in the notation of Section A.2 (see Appendix 5.4 for details):

$$f_i(t) = \frac{212\sqrt{2}}{(iII)}[1.0-(-1.0)^i], \tag{5.115}$$

$$z_{1i}(t) = \sqrt{2}(i\Pi)\{-[70+10\sin(2\Pi)t+k_1N_1(t)] + (-1.0)^i[54.5+k_2N_2(t)]\}, \tag{5.116}$$

$$z_{2i} = \frac{70\sqrt{2}}{(i\Pi)}[1.0-e^{-0.25}(-1.0)^i] + \frac{k_3\sqrt{2}}{(i\Pi)}[1.0-(-1.0)^i], \tag{5.1 17}$$

$$z_i(t) = y_i(t;\mathbf{u}^*) + \frac{k_0N_0(t)\sqrt{2}}{(i\Pi)}[1.0-(-1.0)^i], \tag{5.118}$$

$$y_i(t;\mathbf{u}^*) = \frac{70\sqrt{2}}{(i\Pi)}[1.0-e^{-0,25}(-1.0)^i]e^{-(i\Pi)^2t} \tag{5.119}$$

$$+\left\{\frac{212\sqrt{2}}{(i\Pi)^3}[1.0-(-1.0)^i] + \frac{70\sqrt{2}}{(i\Pi)} - \frac{54.5(-1.0)^i\sqrt{2}}{(i\Pi)}\right\}\{1.0-e^{-(i\Pi)^2t}\}$$

$$+\frac{10\sqrt{2}(i\Pi)}{(i\Pi)^4+4\Pi^2}[(i\Pi)^2\sin 2\Pi t - 2\Pi\cos 2\Pi t + 2\Pi e^{-(i\Pi)^2t}], \tag{5.126}$$

$$\{WW^T\}_{k,l} = 2\Pi^2kl[1.0+(-1)^k(-1)^l]. \tag{5.121}$$

Using the identities (5.100)–(5.121), we obtained results for the following, via the developments of Section A.1.1:

(a) $\hat{y}_i(t)$, $\{P_{ij}(t)\}^{-1}$ $(i,j=1,2,3,4)$ evaluated for $(t=0+\Delta t, 0+2\Delta t,\ldots, 1.0)$. $\Delta t = 0.001$, from (5.66), (5.67), (5.68), and (5.69).

(b) $\hat{y}_m(x,t)$ from (5.113), with $(x=0+\Delta x,\ 0+2\Delta x,\ldots,1)$, $\Delta x = 0.01$.

(c) Using the algorithm of Section A.1.2, we recovered $u_1(0,t)$, $u_1(1,t)$ for $(t=0+\Delta t,\ 0+2\Delta t,\ldots,T)$, $\Delta t = 0.01$.

(d) $u_2(x)$ for $(x = 0+\Delta x, 0+2\Delta x, \cdots, 1)$, $\Delta x = 0.01$.
(e) $y_i(t;\mathbf{u})$ for $(t = 0+\Delta t, 0+2\Delta t, \cdots, 1)$, $\Delta t = 0.01$.
(f) $y_m(x, t;\mathbf{u})$ for $x = (0+\Delta x, 0+2\Delta x, \cdots, 1)$, $\Delta x = 0.01$.

A discussion on these results is delayed until after consideration of the second numerical example, in order that comparisons may be made.

Example 2

The system S_D is the same as that of Example 1, evolving according to (5.88)–(5.91). The input measurement are also the same as those in Example 1 given by (5.92)–(5.97), As we indicated earlier, the point of departure was in the definition of the output measurements:

Output Measurements O. Measurements were taken at four points in the spatial domain, for each t. The four points were chosen at $x = 0.2$, $x = 0.4$, $x = 0.6$, $x = 0.8$. That is,

$$z(x^i, t) = y(x^i, t;\mathbf{u}^*) + k_0{}^i N_0{}^i(t), \qquad i = 1, 2, 3, 4)\,, \tag{5.122}$$
$$x^1 = 0.2, \quad x^2 = 0.4, \quad x^3 = 0.6, \quad x^4 = 0.8\,.$$

$y(x^i, t;\mathbf{u}^*)$ is the response of S_D to initial and boundary conditions $\mathbf{u}^*$, defined by (5.95)–(5.97), at the selected points x^i. For each i, $k_0{}^i$ and $N_0{}^i(t)$ are as defined in Example 1.

As before, the problem is to choose $\mathbf{u} \in V$ which extremizes an error functional $J(\mathbf{v})$, $\mathbf{v} \in V$. Here we define the following:

$$V = L^2(\Sigma) \times L^2(\Omega)\,,$$

$$J(\mathbf{v}) = \int_0^1 \sum_{i=1}^{4} [y(x^i, t;\mathbf{v}) - z(x^i, t)]^2\, dt$$
$$+ \int_0^1 [v_1(0,t) - z_1(0,t)]^2\, dt + \int_0^1 [v_1(1,t) - z_1(1,t)]^2\, dt$$
$$+ \int_0^1 [v_2(x) - z_2(x)]^2\, dx\,. \tag{5.123}$$

Remark. The output measurements O for Example 2 are taken at selected points $x^i \in \Omega$. Supposedly, we have access to the system only at the points x^i. Note, however, that one of the "input" measurements, namely $z_2(x)$, is assumed to be available over the entire spatial profile, an apparent inconsistency. First, let it be said that the assumption that $z_2(x)$ is available for all $x \in \Omega$ is essential to the theoretical framework. That framework is, in abstract, the minimization of quadratic functionals on a Hilbert space V. As we saw, V was chosen to be $L^2(\Sigma)\times L^2(\Omega)$. Discrete spatial measurements on the boundary would lead to spaces V such as:

$$V = L^2(\Sigma)\times 1_1{}^2 \times 1_1{}^2 \times \dots \times 1_v{}^2 ,$$

where 1_i^2 is the space of squared real numbers. No consideration of this type space is made. In fact, such a consideration would be extraordinarily difficult. However, we can rationalize this apparent inconsistency. We do so by asserting that $z_2(x)$, the initial "measurement" of the true state of nature $u_2{}^*(x)$ is obtained by *calculation of a steady state profile*. This calculation is assumed to be in error by an amount k_3 (see (5.94)). Thus we assume, for the purposes of this example, that the system S_D is initially at some steady state. This steady state is calculated and used as an initial condition $z_2(x)$, considered as an "input measurement."

The solution of Example 2 is accomplished via Theorem 4.1,IB, which asserts that the unique $\mathbf{u}\in V$ which extremizes $J(\mathbf{v})$ (equation 5.123) is characterized by the following system of equations:

$$\frac{\partial y(x,t;\mathbf{u})}{\partial t} - \frac{\partial^2 y(x,t;\mathbf{u})}{\partial x^2} = 212.0, \qquad x,t\in(0,1)\times(0,1], \tag{5.124}$$

$$y(0,t) = u_1(0,t), \qquad t\in(0,1], \tag{5.125}$$

$$y(1,t) = u_1(1,t), \qquad t\in(0,1], \tag{5.126}$$

$$y(x,0) = u_2(x), \qquad x\in(0,1). \tag{5.127}$$

$$-\frac{\partial p(x,t;\mathbf{u})}{\partial t} - \frac{\partial^2 p}{\partial x^2} = \sum_{i=1}^{4} [y(x^i,t;\mathbf{u}) - z(x^i,t)]\delta(x-x^i), \tag{5.128}$$

$$x,t\in(0,1)\times(0,1], \tag{5.129}$$

$$p(0,t) = 0, \qquad t\in(0,1], \tag{5.130}$$

$$p(1,t) = 0, \qquad t\in(0,1], \tag{5.131}$$

$$p(x,1) = 0, \qquad x\in(0,1). \tag{5.132}$$

$$\frac{\partial p(0,t;\mathbf{u})}{\partial x} + u_1(0,t) - z_1(0,t) = 0, \qquad t\in(0,1], \tag{5.133}$$

$$-\frac{\partial p(1,t;\mathbf{u})}{\partial x} + u_1(1,t) - z_1(1,t) = 0, \qquad t\in(0,1], \tag{5.134}$$

$$p(x,0;\mathbf{u}) + u_2(x) - z_2(x) = 0, \qquad x\in(0,1). \tag{5.135}$$

As in Example 1, the system state $y(x, t; \mathbf{u})$ and its adjoint $p(x, t; \mathbf{u})$ were approximated by:

$$y_m(x, t; \mathbf{u}) = \sum_{i=1}^{4} y_i(t; \mathbf{u}) w_i(x), \tag{5.136}$$

$$p_m(x, t; \mathbf{u}) = \sum_{i=1}^{4} p_i(t; \mathbf{u}) w_i(x), \tag{5.137}$$

$$\hat{y}_m(x, t) = \sum_{i=1}^{8} \hat{y}_i(t) w_i(x), \tag{5.138}$$

$$w_i(x) = \sqrt{2} \sin(\sqrt{\lambda_i})x, \qquad \lambda_i = (i\Pi)^2. \tag{5.139}$$

In the computation of $\hat{y}_i(t)$, the system of identities (5.115)–(5.121) is augmented by the addition of:

$$Q_v = \sum_{i=1}^{v} Q^i, \qquad v = 1, 2, 3, 4 \qquad \text{(the four measurement locations)},$$

$$\{Q^i\}_{k,l} = \sqrt{2} \sin(k\Pi)x^i \sqrt{2} \sin(l\Pi)x^i, \qquad k, l = 1, 2, \ldots, 8. \tag{5.140}$$

Using the appropriate identities contained in this section, we obtained results, via the developments of Section A.1.1, for the following:

(a) $\hat{y}_i(t)$; $\{P_{ij}(t)\}^{-1}$, $i, j = 1, 2, \cdots, 8$,
(b) $\hat{y}_m(x, t)$ using (5.138),
(c) using the algorithm of Section A.1.2, we recovered $u_1(0, t), u_1(1, t)$, and
(d) $u_2(x)$,
(e) $y_i(t; \mathbf{u})$
(f) $y_m(x, t; \mathbf{u})$.

The intervals of definition of these quantities are the same as in Example 1. We are now in a position to give an evaluation of the numerical results for Examples 1 and 2.

A.4 EVALUATION OF THE NUMERICAL RESULTS

Numerical Examples 1 and 2 are considered in tandem because of the pervading structural similarity of the equations defining the solutions. We consider first solutions for $\hat{y}(x, t)$, the "filtered estimate" for both examples.

The equations of interest are those arising out of an approximation to $\hat{y}(x,t)$ (see Section A.1.1, A.2).

Example 1.

$$\frac{d\hat{\mathbf{y}}(t)}{dt} + \mathbf{A}\hat{\mathbf{y}}(t) - \mathbf{f}(t) + \mathbf{z}_1(t) = \mathbf{P}^{-1}(t)[\mathbf{z}(t) - \hat{\mathbf{y}}(t)], \tag{5.66}$$

$$\hat{\mathbf{y}}(0) = \mathbf{z}_2, \tag{5.67}$$

$$\frac{d\mathbf{P}^{-1}(t)}{dt} + \mathbf{A}\mathbf{P}^{-1}(t) + \mathbf{P}^{-1}(t)\mathbf{A} + \mathbf{P}^{-1}(t)\mathbf{I}\mathbf{P}^{-1}(t) - \mathbf{W}\mathbf{W}^T = 0, \tag{5.64}$$

$$\mathbf{P}^{-1}(0) = \mathbf{I}, \tag{5.65}$$

$$\hat{y}_m(x,t) = \sum_{i=1}^{m} \hat{y}_i(t) w_i(x).$$

Example 2.

$$\frac{d\hat{\mathbf{y}}}{dt} + \mathbf{A}\hat{\mathbf{y}}(t) - \mathbf{f}(t) + \mathbf{z}_1(t) = \mathbf{P}^{-1}(t)\mathbf{Q}_v[\mathbf{z}(t) - \hat{\mathbf{y}}(t)], \tag{5.84}$$

$$\hat{\mathbf{y}}(0) = \mathbf{z}_2, \tag{5.85}$$

$$\frac{d\mathbf{P}^{-1}(t)}{dt} + \mathbf{A}\mathbf{P}^{-1}(t) + \mathbf{P}^{-1}(t)\mathbf{A} + \mathbf{P}^{-1}(t)\mathbf{Q}_v\mathbf{P}^{-1}(t) - \mathbf{W}\mathbf{W}^T = 0, \tag{5.86}$$

$$\mathbf{P}^{-1}(0) = \mathbf{I}, \tag{5.87}$$

$$\hat{y}_m(x,t) = \sum_{i=1}^{m} \hat{y}_i(t) w_i(x).$$

In both examples, a fourth-order Runge-Kutta integration routine was used to generate solutions. It was found that a step size of 0.001 had to be taken to ensure the numerical stability of the (8×8)-dimensional P equations. This was a consequence of the small time constant τ, $\tau = 1/[2(i\Pi)^2]$, for higher-order modes. This small stepsize resulted in the algorithm being slow, although the computation time required for each integration step (manipulating and integrating 52 equations), was about 3 seconds.

Figure 5.1 shows a typical set of responses of elements of $\mathbf{P}^{-1}(t)$. Note the very fast "rise time" of the higher-order mode $\{\mathbf{P}^{-1}\}_{5,3}$. Note also the

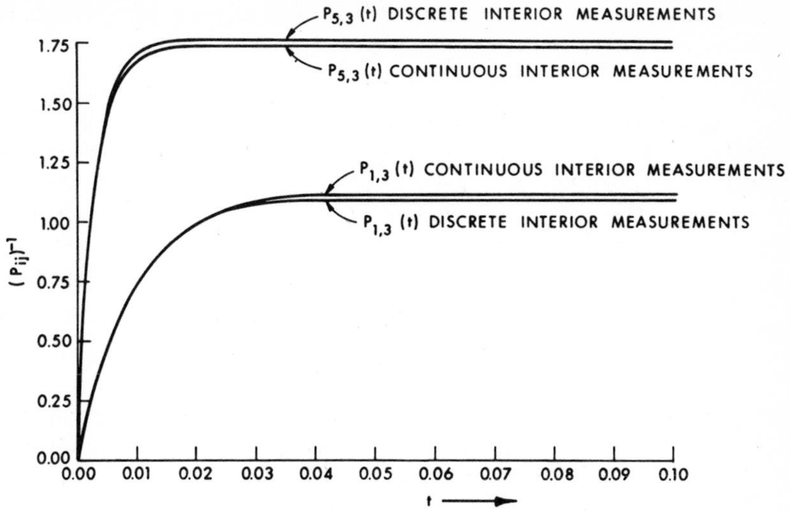

Figure 5.1. Response of selected elements of $(P_{ij})^{-1}$.

offset at steady state between the results for Examples 1 and 2. A positive definite character of $\mathbf{P}^{-1}(t)$ for all t was observed for $t_{ss}=0.008$, for Example 1. The elements of $\mathbf{P}^{-1}$ were at steady state for $t_{ss}=0.008$. The results are displayed in Table 5.1.

TABLE 5.1 Principal Minors of $P^{-1}[t_{ss}]$

Principal minor number	Value
1	$1.7349>0$
2	$3.2860>0$
3	$3.9406>0$
4	$2.7196>0$
5	$0.9916>0$
6	$0.1749>0$
7	$0.0158>0$
8	$0.0050>0$

An important feature of the matrix $\mathbf{P}^{-1}(t)$ in Example 1, for all $t \in (0, T]$, is that:

$$\{P^{-1}(t)\}_{i,j} = 0, \quad \text{if } (i+j) \text{ is odd.}$$

We see this property immediately from the equations defining $\mathbf{P}^{-1}(t)$, wherein it is observed that the odd terms satisfy a homogeneous linear ordinary differential equation with zero initial conditions.This property plays an important role in the definition of the responses $\hat{y}(t)$, to which we now turn.

In Example 1 the equation for the ith component of $\hat{y}(t)$ is (equation 5.66):

$$\frac{d\hat{y}_i(t)}{dt} + (i\Pi)^2 \hat{y}_i(t) + f_i(t) - z_{1i}(t) = \sum_{j=1}^{8} \{P^{-1}(t)\}_{ij}[z_j(t) - \hat{y}_j(t)]. \tag{5.141}$$

Now observe that, i even, z_{2i}, $z_{1i}(t)$, and $z_i(t)$ contain no noise—a consequence of the eigenfunction expansion chosen. Thus the LHS of (5.141) describe exactly the evolution of the ith mode of the true state of nature. The property that $\{\mathbf{P}^{-1}(t)\}_{ij} = 0$ if $(i+j)$ odd implies that the RHS of (5.141) is zero for *i even*. Thus for, i even, (5.141) describes the evolution of the ith mode of the true state of nature.

A similar result holds for the equations of Example 2.

The foregoing discussion is summarized by the response shown for $\hat{y}_2(t)$, Examples 1 and 2, in Figure 5.2. A typical response of an odd-numbered

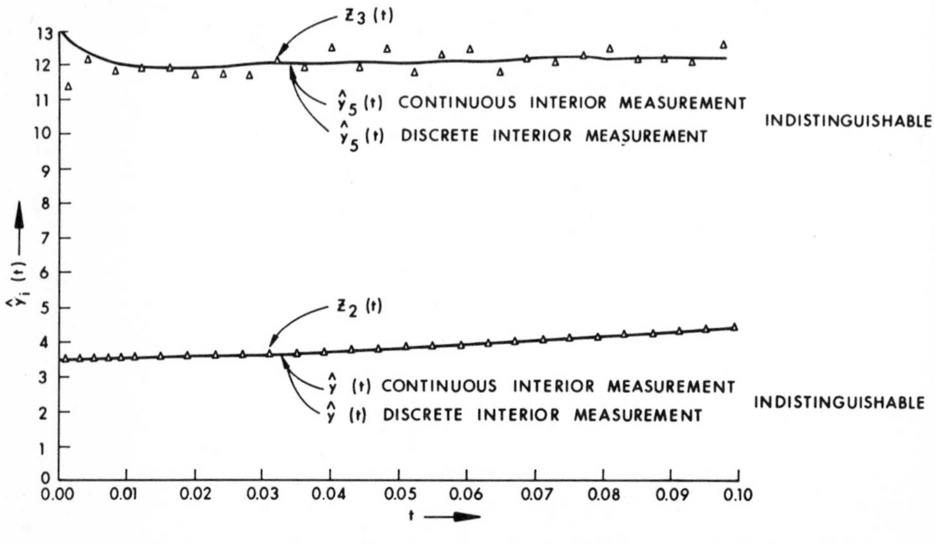

Figure 5.2. Response of selected modes of the filtered estimates $\hat{y}_j(t)$.

element of $\hat{y}(t)$, Examples 1 and 2, is also given in Figure 5.2. It is noted that the responses $\hat{y}_5(t)$ shown in Figure 5.2 for both examples are virtually indistinguishable. However, it is possible to distinguish between the responses $\hat{y}_1(t)$ for the two examples, as shown in Figure 5.3. We observed that:

$$[\hat{y}_i(\cdot)]_{\text{Ex. 2}} \to [\hat{y}(\cdot)]_{\text{Ex. 1}} \qquad \text{as } i \to \infty .$$

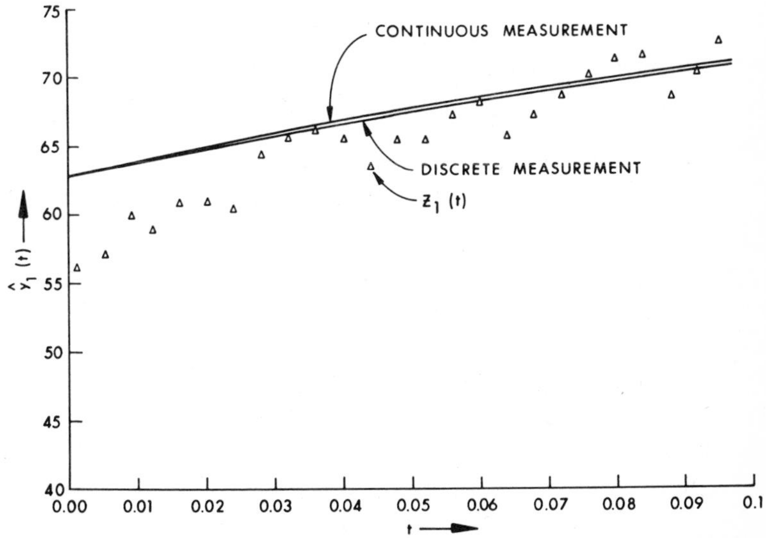

Figure 5.3. Response of $\hat{y}_1(t)$. Discrete and continuous interior measurement.

A graphical display of $\hat{y}_m(x,t)$ (and $P_m(x,\xi,t)$) is omitted, as the salient features of this kind of approximation are treated in the consideration of the approximation to the "smoothed estimate," $y_m(x,t;\mathbf{u})$, which follows. To illustrate the method of solution, we shall go through an iteration of the algorithm for obtaining the "smoothed estimate" $y(x,t;\mathbf{u})$ given in Section A.1.3, using therein the approximations outlined in Section A.2. The specific examples are those of this section.

Iteration 0. (i) Guess the functions $u_1(0,.)$, $u_1(1,.)$, and $u_2(.)$ over the appropriate grid (Section A.3).

Remark. A reasonable first guess is to set

$$u_1{}^0(0,\cdot) = z_1(0,\cdot)\,,$$
$$u_1{}^0(1,\cdot) = z_1(1,\cdot)\,,$$
$$u_2{}^0(\cdot) \;\; = z_2(\cdot)\,.$$

For the purpose of illustrating convergence properties, the first guess of $\mathbf{u}^0$ was made independently of the measurements $\mathbf{z}$. See, for example, the first guess of $u_2{}^0(\cdot)$ illustrated in Figure 5.5.

(ii) Evaluate $\mathbf{G}(\mathbf{u}^0)$, given by:

$$\mathbf{G}(\mathbf{u}^0) = \begin{array}{l} \sum_{i=1}^{4} p_i(.)[\sqrt{2}(i\Pi)\cos(i\Pi)x]_{x=0} + u_1{}^0(0,.) - z_1(0,.) \\ -\sum_{i=1}^{4} p_i(\cdot)[\sqrt{2}(i\Pi)\cos(i\Pi)x]_{x=1} + u_1{}^0(1,\cdot) - z_1(1,\cdot) \\ \sum_{i=1}^{4} p_i(0)\sqrt{2}\sin(i\Pi)(\cdot) + u_2{}^0(\cdot) - z_2(\cdot) \end{array} . \tag{5.89}$$

In order to carry out the evaluation, we must first obtain $p_i(\cdot)$ and $p_i(0)$, $(i = 1,2,3,4)$. $p_i(\cdot)$ and $p_i(0)$ are obtained by integrating "backward" in time (from $t = 1$) the following equations:

$$-\frac{dp_i(t)}{dt} + (i\Pi)^2 p_i(t) = y_i(t;\mathbf{u}^0) - z_i(t), \tag{5.90}$$

$$p_i(1) = 0. \tag{5.91}$$

$y_i(\cdot;\mathbf{u}^0)$ is obtained by solving (from $t = 0$):

$$\frac{dy_i(t)}{dt} + (i\Pi)^2 y_i(t) = f_i(t) - u_{1i}^0(t), \tag{5.92}$$

$$y_i(0) = u_{2i}^0, \tag{5.93}$$

where

$$u_{1i}^0(t) = \sqrt{2}(i\Pi)\{-[u_1{}^0(0,t)] + (-1)^i[u_1{}^0(1,t)]\},$$

$$u_{2i} = \int_0^1 \sqrt{2}u_2{}^0(x)\sin(i\Pi)x\,dx.$$

Remark. It was found that an integration step size of 0.01 was sufficient to guarantee the numerical stability of a fourth-order Runge-Kutta integration scheme.

Having solved for $p_i(t)$, $t \in (0,1]$ from (5.90) through (5.93), $\mathbf{G}(\mathbf{u}^0)$ is evaluated via (5.89).

(iii) If $\mathbf{G}(\mathbf{u}^0)$ is nonzero (as it would be except by most propitious circumstance), then the choice of $\mathbf{u}^0$ is updated according to:

$$u_1{}^1(0, .) = u_1{}^0(0, \cdot)+\alpha^0 S_1{}^0(0, \cdot),$$
$$u_1{}^1(1, \cdot) = u_1{}^0(1, \cdot)+\alpha^0 S_1{}^0(1, \cdot),$$
$$u_2{}^1(\cdot) \quad = u_2{}^0(\cdot) \quad +\alpha^0 S_2{}^0(\cdot).$$

As we indicated (Section A.1.3), $\mathbf{S}^0$ is chosen to be the negative gradient which has already been computed:

$$S_1^0(0,\cdot) = -G_1^0(0,\cdot) = -\sum_{i=1}^{4} p_i(\cdot)[\sqrt{2}(i\Pi)\cos(i\Pi)x]_{x=0} - u_1{}^0(0,\cdot)+z_1(0,\cdot),$$

$$S_1^0(1,\cdot) = -G_1^0(1,\cdot) = \sum_{i=1}^{4} p_i(\cdot)[\sqrt{2}(i\Pi)\cos(i\Pi)x]_{x=1} - u_1{}^0(1,\cdot)+z_1(1,\cdot),$$

$$S_2{}^0(\cdot) = -G_2{}^0(\cdot) = -\sum_{i=1}^{4} p_i(0)\sqrt{2}\sin(i\Pi)(.) - z_2{}^0(\cdot)+z_2(\cdot).$$

α^0 is given explicitly by:

$$\alpha^0 = \frac{\int_0^1 \{G_1{}^0(0,t)^2 + G_1{}^0(1,t)^2\}\,dt + \int_0^1 G_2{}^0(x)^2\,dx}{\sum_{i=1}^{4}\int_0^1 [y_i(t;\mathbf{S}^0) - y_i(t;\mathbf{0})]^2\,dt + \int_0^1 \{S_1{}^0(0,t)^2 + S_1{}^0(1,t)^2\}\,dt + \int_0^1 S_2{}^1(x)^2\,dx}.$$

Thus we obtain the $\mathbf{u}^1$ vector. The computational effort is less forbidding than the algebra implies.

Successive stages in the algorithm are as outlined in section A.1.3. The appropriate modifications to the explicit formulas given for the "zeroth" iteration are straightforward.

With this understanding of the iterative scheme, we can proceed to a discussion of its application to the two examples of this section. We obtained the following results, common to both examples:

(i) Convergence of the algorithm is rapid: 3 iterations sufficed for Example 1, and 4 iterations for Example 2 (that is, $\mathbf{G}(\mathbf{u}^3) = 0$ and $\mathbf{G}(\mathbf{u}^4) = 0$, respectively). Computation time per iteration was 25 seconds.

(ii) Using the refined estimate $\mathbf{u}$ obtained after the appropriate number of iterations, the approximation to $p(x,t;\mathbf{u})$ afforded by the four-mode expansion

$$p_m(x,t;\mathbf{u}) = \sum_{i=1}^{4} p_i(t;\mathbf{u})w_i(x)$$

was precise, $p_1(t;\mathbf{u})$ being the only nonzero mode in the expansion. That is,

$$\lim_{m\to\infty} p_m(x,t;\mathbf{u}) = p_1(t;u)w_1(x) = p(x,t;\mathbf{u})$$

for almost every $x, t \in (0,1)\times(0,1]$.

A consequence of this result was that $\mathbf{u}$, defined in terms of $p(x,t;\mathbf{u})$, was accurate.

(iii) Whereas a four-mode expansion was adequate to define $p(x,t;\mathbf{u})$, the approximation of $y_m(x,t;\mathbf{u})$ to $y(x,t;\mathbf{u})$ was poor as will be seen. This fact is of little consequence, since the announced goal was to find $\mathbf{u}$, not $y(x,t;\mathbf{u})$.

Figures 5.4 through 5.15 illustrate the numerical results for Examples 1 and 2. In Figure 5.4 we show the refined estimate of one of the boundary conditions. The convergence of the first guess of the other boundary condition $u_1{}^0(1,t)$ is displayed in Figure 5.5. In Figure 5.6 we show the refined estimate

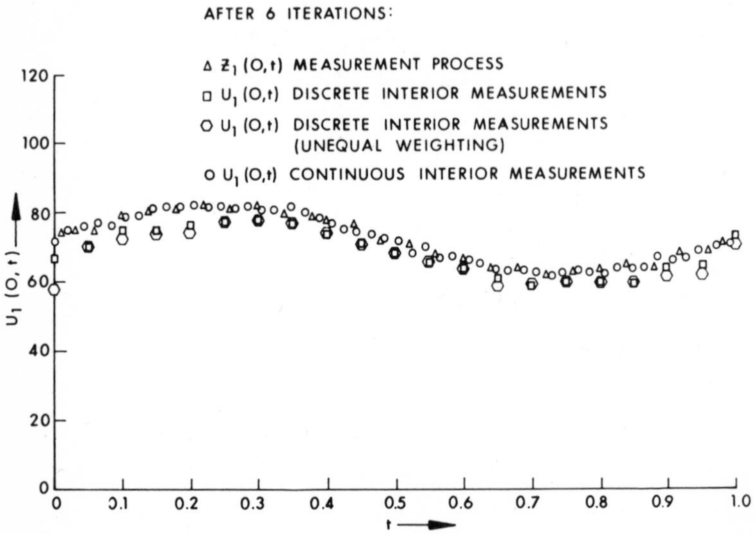

Figure 5.4. Boundary condition $u_1(O, t)$.

of the initial condition. Figures 5.7 and 5.8 show the responses of typical modes of the refined estimate $y_i(t; \mathbf{u})$ and the effect of successive improvement in the choice of $\mathbf{u}$.

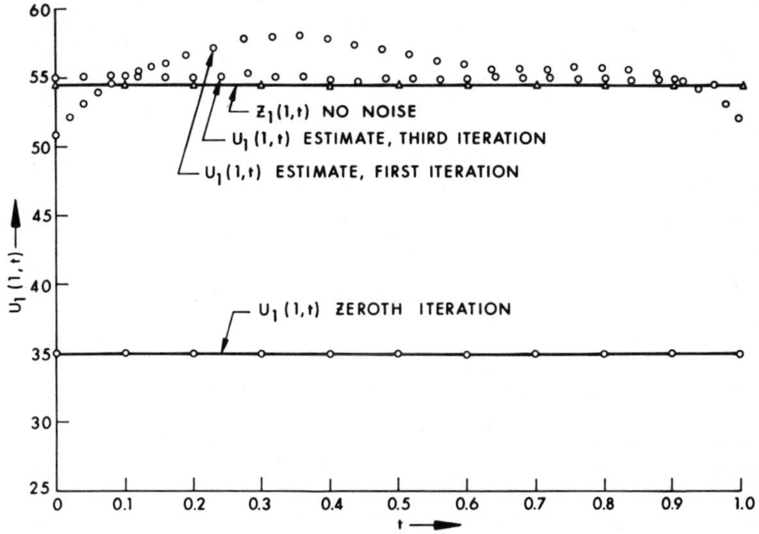

Figure 5.5. Boundary condition $u_1(1, t)$.

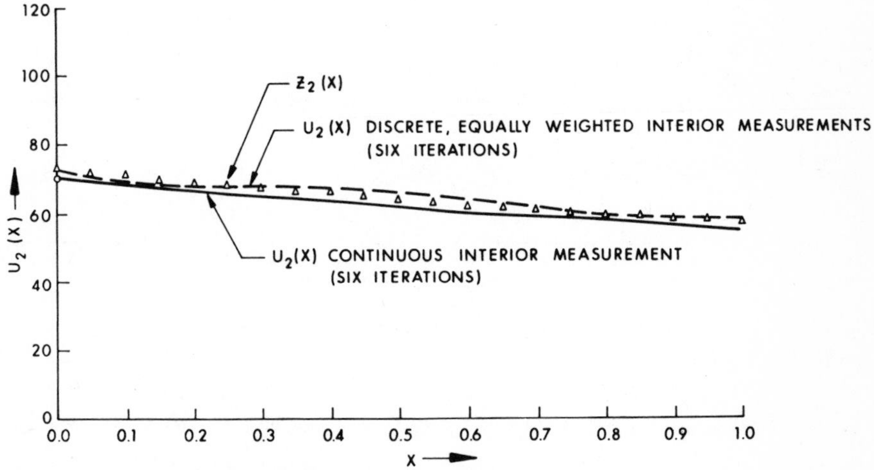

Figure 5.6. Initial condition $u_2(X)$.

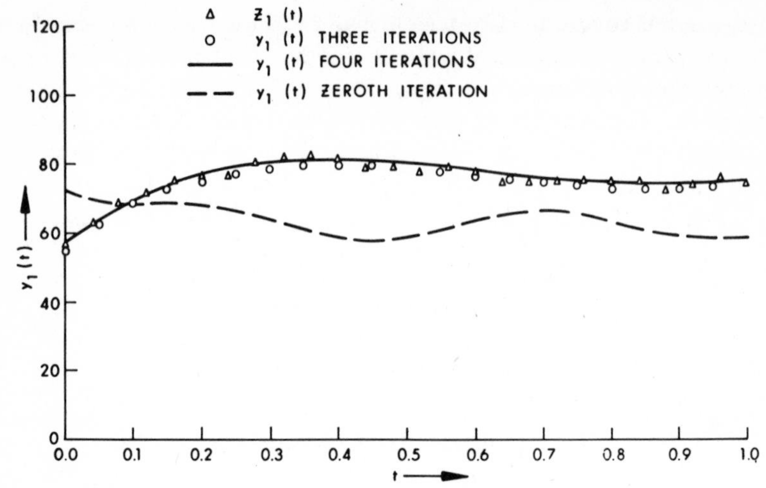

Figure 5.7. Response of the first mode of the refined estimate, $y_1(t)$.

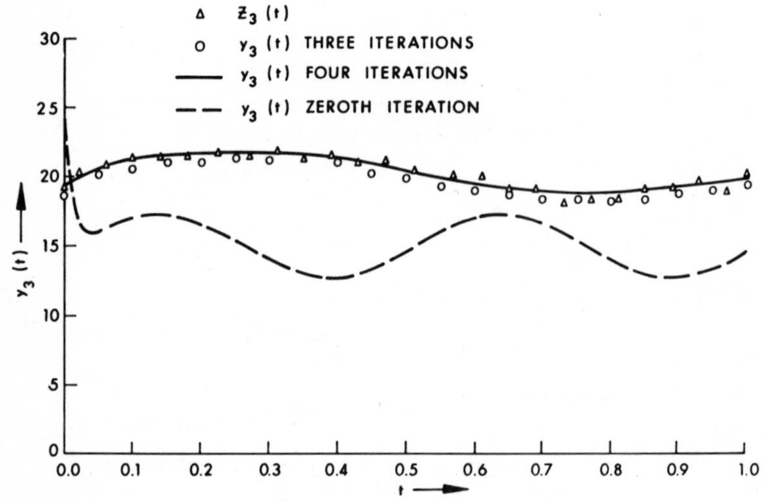

Figure 5.8. Response of the third mode of the refined estimate $y_3(t)$.

Using $y_i(t;\mathbf{u})$, we recover the approximation to $y(x,t;\mathbf{u})$ as follows:

$$y_m(x,t;\mathbf{u}) = \sum_{i=1}^{4} y_i(t;\mathbf{u})\, h_i(x).$$

This approximation is shown in Figure 5.9 for $t = 0.4$. We note that the approximation is singularly bad. Of course, the addition of more terms in the approximation would improve the fit but, in the light of preceding arguments, the accuracy of **u** would not be improved, note that $y_m(x, t; \mathbf{u})$ could be obtained to any desired accuracy, following the recovery of **u** with a four-mode model.

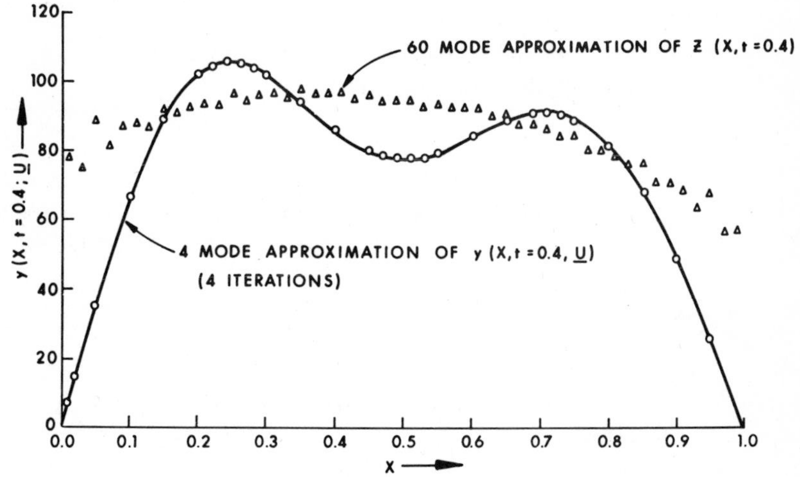

Figure 5.9. Spatial profile of refined estimate at $t = 0.4$.

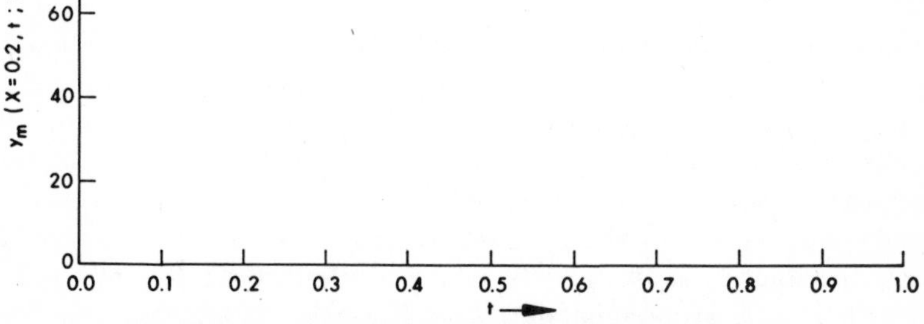

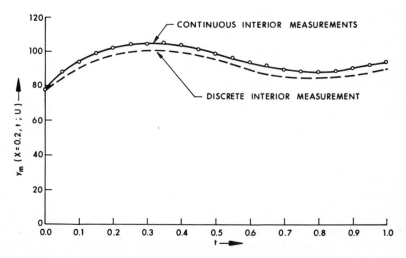

Figure 5.10. Comparison of four mode refined estimates.

A comparison of the four-mode refined estimates for Examples 1 and 2 is afforded by Figure 5.10, wherein we show $y(0.2, t; u)$ for examples 1 and 2. Note from Figure 5.9 that at $x = 0.6$ the four-mode approximations fall below the measured data. We show in Figure 5.11 the response at $x = 0.6$ of the four-mode refined estimate for Example 2. By "equal weighting" and "unequal weighting" we mean that the measurement errors at the discrete spatial locations were equally or unequally weighted. Unequal weighting caused a deterioration in the accuracy of the refined estimate. The apparent improvement shown in Figure 5.11 is illusory, since the fit was inordinatly worsened at the other spatial locations.

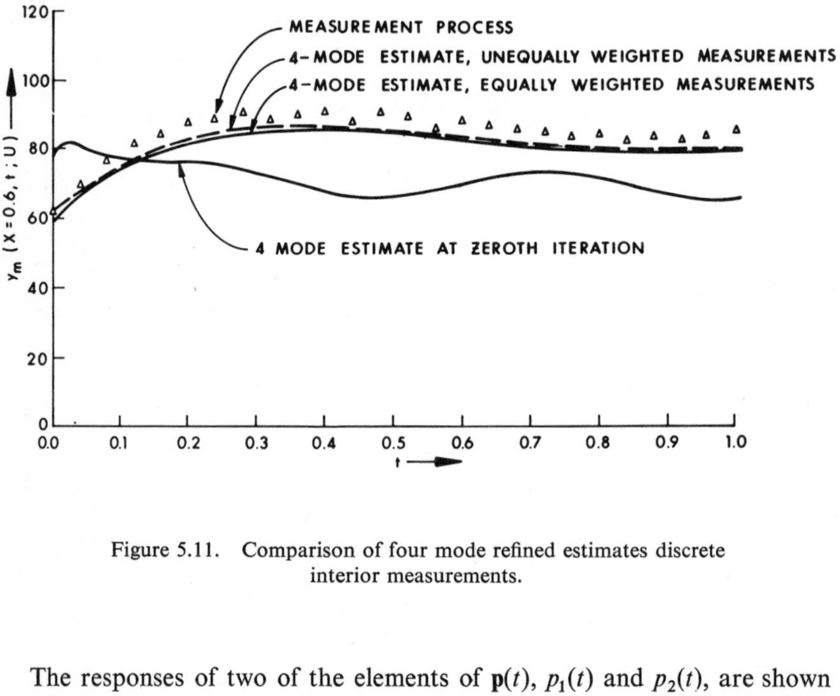

Figure 5.11. Comparison of four mode refined estimates discrete interior measurements.

The responses of two of the elements of $\mathbf{p}(t)$, $p_1(t)$ and $p_2(t)$, are shown in Figure 5.12 Note the effect of successive iterations on the responses. We observed that, after the third iteration, $p_2(.)$, $p_3(\cdot)$, and $p_4(\cdot)$ were all identically zero. $p_1(\cdot)$ was nonzero as shown. An indication of the rate of convergence of the algorithm is shown in Figures 5.13 and 5.14. In the former the convergence of one of the gradients of the functional $J(\mathbf{u})$ to zero is shown. In the latter we show the reduction of $J(\mathbf{u})$ with successive iterations. Note that the value of the minimum J for Example 1 is less than that for Example 2.

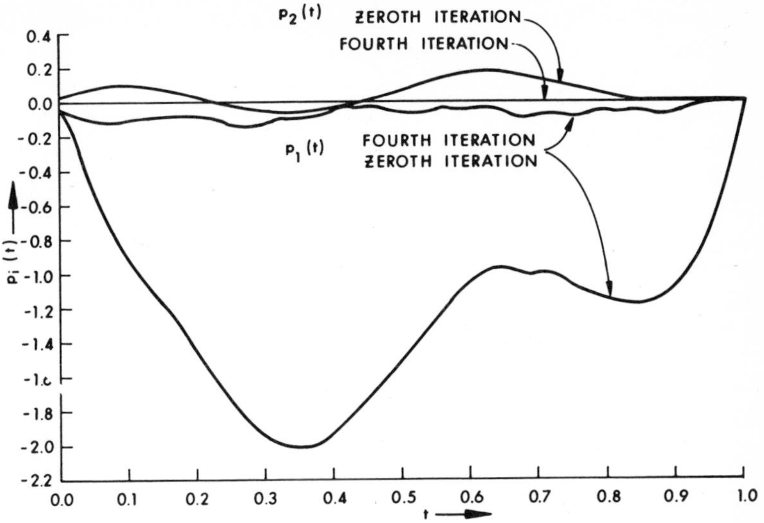

Figure 5.12. Response of selected modes of the adjoint variable $p_i(t)$.

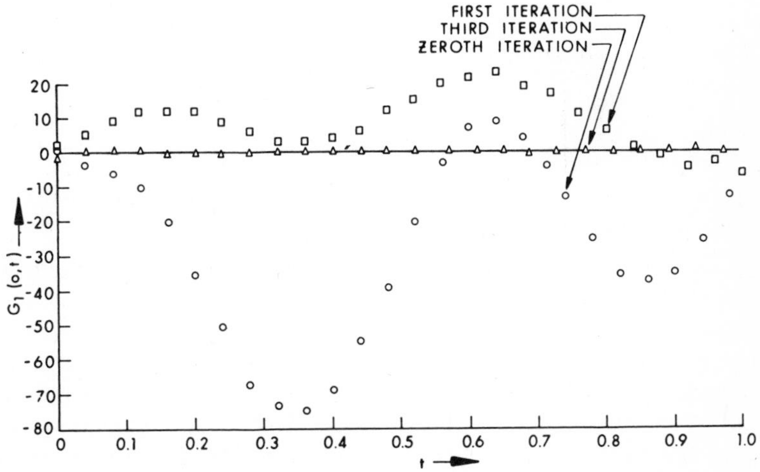

Figure 5.13. Response of a gradient of the error functional.

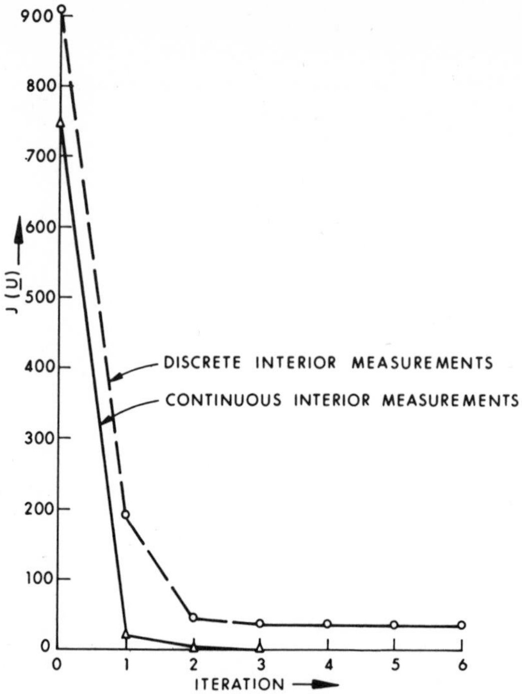

Figure 5.14. Evaluation of the error functional.

Some general remarks concerning the simulation are in order:

(i) No significant loss in the accuracy of the refined boundary and initial estimates was incurred by taking measurements at discrete interior spatial locations. This is not an obvious result and it cannot be generalized, though it may indicate that some of the information contained in the continuous spatial measurements is redundant.

(ii) The computational efficiency of the successive approximation algorithm presented should be stressed. Rapid convergence and relative computational speed per iteration of the scheme are noteworthy. A solution for the optimal **u** was obtained in 75 seconds on a Univac 1107. The computing time could undoubtedly be reduced by the use of more sophisticated program ming. By comparison, obtaining an eight-mode approximation of the filtered estimate, $\hat{y}_m(x, t)$ would take 3000 seconds. Of course, the filtered estimate is a sequential or on-line estimator, whereas the smoothed estimate is obtained off-line ,so direct comparisons are not meaningful.

A.5 SUMMARY

In summary, this chapter has dealt with two distributed identification problems from formulation to numerical evaluation of two special examples. Although the numerical evaluation was relatively exhaustive, there remain several alternatives for the numerical approximation of the solutions to the partial differential equations. This consideration, along with others, is proposed as meaningful extensions in the next chapter.

CHAPTER 6

SUMMARY AND EXTENSIONS

A.1 THE PARAMETER IDENTIFICATION PROBLEM

The state identification problem associated with various parabolic and hyperbolic systems was treated in some detail. Of fundamental importance was the familar "linear system, quadratic cost" setting for, in that case, existence and uniqueness of solutions were demonstrated under a few weak assumptions. In addition, the now classical Kalman-Bucy type of equations emerged, reinforcing our intuitive feelings about the connection between lumped and distributed systems.

An important asset of the variational approach which was taken here is that direct computational methods suggested themselves; that is, computational algorithms of the "successive sweep" type arose quite naturally in this context. Of significant importance is the fact that these algorithms do not rely on the linearity of the system or the quadratic nature of the "cost" functional, but are generally applicable. Of course, this computational flexibility is semisweet, since the existence and uniqueness properties are either absent or are present at the expense of strong assumptions elsewhere. As an example of the sort of compromise which can be made in the solution of nonlinear estimation problems, we consider, as an extension of the techniques already presented, a special formulation of a parameter identification problem. Consider, then, the following:

State evolution process:

$$\frac{\partial y(x,t)}{\partial t} - a(x)\frac{\partial^2 y(x,t)}{\partial x^2} = f(x,t) \quad \text{in } Q, \tag{6.1}$$

$$y(\Sigma) = u_1(\Sigma) \quad \text{on } \Sigma, \tag{6.2}$$

$$y(x,0) = u_2(x) \quad \text{in } \Omega. \tag{6.3}$$

As before, $u_1(\Sigma)$ and $u_2(x)$ are unknown, and are to be determined by minimization of an appropriate quadratic error functional. Suppose that, in addition, $a(x)$ is unknown, except for an initial estimate, given by

$$[a(x)]_{\text{initial estimate}} = u_3(x). \tag{6.4}$$

We reformulate the system as follows:

Define

$$y_1(x,t) = y(x,t), \qquad x,t \in Q,$$

$$y_2(x,t) = a(x), \qquad x,t \in Q.$$

Then the system (6.1), (6.2), (6.3), and (6.4) may be written:

$$\frac{\partial y_1(x,t)}{\partial t} - y_2(x,t)\frac{\partial^2 y_1(x,t)}{\partial x^2} = f(x,t), \qquad \frac{\partial y_2(x,t)}{\partial t} = 0, \tag{6.5}$$

with the initial and boundary conditions:

$$y_1(x,0) = u_2(x), \tag{6.6}$$

$$y_2(x,0) = u_3(x), \tag{6.7}$$

$$y_1(\Sigma) = u_1(\Sigma). \tag{6.8}$$

We note that equations given by (6.5) are not of the type treated earlier. We can, however, reduce (6.5) to the type considered by carrying out two approximations.

(I) Linearize (6.5) about an initial guess $\bar{y}_1(x,t)$ and $\bar{y}_2(x,t)$. Obtain equations for the principal-linear part of the perturbations, $p_1(x,t)$ and $p_2(x,t)$, induced perturbing $u_1(\Sigma)$, $u_2(x)$, and $u_3(x)$:

$$y_1(x,t) = \bar{y}_1(x,t) + p_1(x,t),$$

$$y_2(x,t) = \bar{y}_2(x,t) + p_2(x,t),$$

$$u_1(\Sigma) = \bar{u}_1(\Sigma) + \delta u_1(\Sigma): \quad u_2(x) = \bar{u}_2(x) + \delta u_2(x); \quad u_3(x) = \bar{u}_3(x) + \delta u_3(x).$$

$$\frac{\partial p_1(x,t)}{\partial t} - c_1(x,t)\,p_2(x,t) - c_2(x,t)\frac{\partial^2 p_1}{\partial x^2} = 0, \tag{6.9}$$

$$\frac{\partial p_2}{\partial t} = 0, \tag{6.10}$$

$$p_1(\Sigma) = \delta u_1(\Sigma); \quad p_2(\Sigma) = 0, \tag{6.11}$$

$$p_1(x,0) = \delta u_2(x); \quad p_2(x) = \delta u_3(x), \tag{6.12}$$

where

$$c_1(x,t) = \frac{\partial^2 \bar{y}_1}{\partial x^2}(x,t)\,,$$

$$c_2(x,t) = \bar{y}_2(x,t)\,.$$

The second approximation necessary would be to introduce, in the manner suggested by Lions and Lattes [24] diffusion terms into (6.9) and (6.10):

(II) Modify (6.9) and (6.10) as follows:

$$\frac{\partial p_1(x,t)}{\partial t} - c_1(x,t)\,p_2(x,t) - \varepsilon_1 \frac{\partial^2 p_1(x,t)}{\partial x^2} - c_2(x,t)\frac{\partial^2 p_1(x,t)}{\partial x^2} = 0\,, \tag{6.13}$$

$$\frac{\partial p_2(x,t)}{\partial t} - \varepsilon_2 \frac{\partial^2 p_2(x,t)}{\partial x^2} = 0\,, \tag{6.14}$$

where ε_1 and ε_2 are allowed to go to zero (in the limit).

Evidently (6.13) and (6.14) are now in the appropriate form. The object of the parameter identification problem would be to choose $\delta u_1(\Sigma)$, $\delta u_2(x)$, $\delta u_3(x)$ which minimize an error functional constrained by (6.13) and (6.14). The optimal choice of $\delta u_3(x)$ yields a solution to the parameter identification problem.

We feel, intuitively, that, given an initial nominal guess of the vector $\mathbf{u}$ close to the optimal solution, the linearization would be effective. Unfortunately, proceeding formally as suggested, the problem loses its mathematical rigor, a cost often incurred but willingly accepted in many practical applications.

Another approach could consist in computing the gradients $G(\mathbf{u})$ (without linearizing) and proceeding via some sucessive sweep algorithm. The only real objection to this approach is the difficulty in computing an appropriate step size, α. A closed-form solution for α is no longer possible, and some direct search technique must be adopted for its selection. So it appears that computational burden is once more conserved.

Another aspects of the problem rich with possibilities but fraught with difficulty is that of numerical technique. Given that a Galerkin-like technique is adopted, the choice of "basis functions" $w_i(x)$ is virtually limitless, and it is not obvious which is most acceptable from the standpoint of accuracy and/or computational effort. We suspect that the choice should be based on:

(i) the type of partial differential equation,
(ii) a compromise between accuracy and computational effort.

Finally, the solution of the integral equation characterizing the filtered estimate might be attempted by one of the two conventional techniques of successive substitution or successive approximation. We suggest this expedient because of the large amount of time required to solved the distributed Ricatti-like equation.

APPENDIX 2.1

DISTRIBUTION DERIVATIVE, EXAMPLE

Consider obtaining the derivative of $f(x)$ defined as follows:

$$f(x) = \begin{cases} a_0, & 0 \le x < x_0\,, \\ a_1, & x_0 < x \le 1\,, \end{cases}$$

$$\int_0^1 f(x)^2\,dx = (a_0{}^2 - a_1)x_0 + a_1{}^2 < \infty\,.$$

Note that $f(x)$ is discontinuous at $x = x_0$ and that df/dx is not, in the usual sense, defined. However, it is possible to recover an expression for df/dx in the sense of Definition 2.1. We have:

$$\int_0^1 \frac{df}{dx}\,\phi(x)\,dx = -\int_0^1 f(x)\frac{d\phi(x)}{dx}\,dx$$

$$= -\left\{\int_0^{x_0^-} a_0\frac{d\phi(x)}{dx}\,dx + \int_{x_0^+}^1 a_1\frac{d\phi}{dx}\,dx\right\} = a_1\phi(x_0{}^+) - a_0\phi(x_0{}^-)\,.$$

But $\phi(x)$ is continuous, so

$$\int_0^1 \frac{df}{dx}\,\phi(x)\,dx = (a_1 - a_0)\,\phi(x_0) = \int_0^1 [(a_1 - a_0)\,\delta(x - x_0)]\,\phi(x)\,dx\,.$$

Collecting terms and once more noting the continuity of $\phi(x)$ and also its arbitrariness, we obtain that

$$\frac{\partial f(x)}{\partial x} = (a_1 - a_0)\,\delta(x - x_0)\,, \tag{i}$$

where $\delta(x - x_0)$ is the Dirac delta function.

APPENDIX 2.2

ERROR ESTIMATES OF GALERKIN APPROXIMATION WITH EIGENFUNCTIONS

The degree of approximation, $E_m(y)$, occasioned by a projection of the solution to system I on the finite dimensional subspace spanned by $\{w_i(x)\}_{i=1,2,\dots,m}$ is taken to be:

$$E_m(y_m) = \min_{a_i(t)} \{\|y(.,t) - \sum_{i=1}^{m} a_i(t) w_i(.)\|_{L^2(\Omega)}\} . \qquad \text{(i)}$$

For $y(\cdot,t) \in L^2(\Omega)$ and $\{w_i\}_{i=1,2,\dots}$ a complete orthonormal system closed in $L^2(\Omega)$, then (i) is equivalent to:

$$E_m^2(y_m) = \|\sum_{i=1}^{\infty} y_i(t) w_i(\cdot) - \sum_{i=1}^{m} y_i(t) w_i(\cdot)\|^2_{L^2(\Omega)}$$

$$= \sum_{i=m+1}^{\infty} y_i^2(t) \qquad (y_i(t) \text{ defined by (2.29)}) \qquad \text{(ii)}$$

A solution to (2.29) is:

$$y_i(t) = e^{-\lambda_i t} u_{2i} + \int_0^t e^{-\lambda_i(t-\tau)} [f_i(\tau) - u_{1i}(\tau)] \, d\tau .$$

Define

$$M = \sup_{i} \sup_{t \in (0,T]} \{|f_i(\cdot) - u_{1i}(\cdot)|\} .$$

Then

$$|\bar{y}_i(t)| \leq \int_0^t \left| e^{-\lambda_i(t-\tau)} (f_i(\tau) - u_{1i}(\tau)) \right| d\tau ,$$

where $\bar{y}_i(t) = y_i(t) - u_{2i}$; thus

$$|\bar{y}_i(t)| \leq M \int_0^t e^{-\lambda_i(t-\tau)} d\tau = \frac{M}{\lambda_i} [1.0 - e^{-\lambda_i t}]$$

Then

$$E_m^{\,2}(\bar{y}_m) \leq \sum_{i=m+1}^{\infty} \left(\frac{M}{\lambda_i}\right)^2 . \tag{iii}$$

A closed form-representation for (iii) is possible, using the well-known formula for monotone decreasing series:

$$\sum_{i=m+1}^{\infty} u_i \leq \int_m^{\infty} f(x)\,dx . \tag{iv}$$

Example.

$$w_i(x) = \sqrt{2}\sin\sqrt{\lambda_i}\,x, \qquad \lambda_i = (i\Pi)^2 .$$

Then

$$E_m^{\,2}(\bar{y}_m) \leq \sum_{i=m+1}^{\infty} \left(\frac{M}{\lambda_i}\right)^2 = \sum_{i=m+1}^{\infty} \frac{M^2}{(i\Pi)^4}$$

$$\leq \int_m^{\infty} \left(\frac{M^2}{\Pi^4}\right) \frac{1}{x^4}\,dx = \left(\frac{M^2}{3\Pi^4}\right) \frac{1}{m^3} . \tag{v}$$

Equation (v) is the result quoted in Section A.3.

APPENDIX 2.3

SPLINE APPROXIMATION OF SYSTEM I

The spline function is, in general, a piecewise polynomial function on a mesh Δ: $a = x_0 \leq x_1 \leq \ldots \leq x_N = b$. The spline function considered here is the cubic spline. For a complete description of the properties of these splines, see [25].

As an example, we shall consider an application of the cubic spline interpolation to the solution of system I with the following data:

$$\Omega = \{x: 0 < x < 1\},$$
$$\Gamma = \{x: x = 0 \text{ and } x = 1\},$$
$$Q = \Omega \times (0, T],$$
$$\Sigma = \Gamma \times (0, T].$$

We shall assume *formally* that $y(.,.) \in C^2(Q)$. Introduce the cardinal splines $\{w_{\Delta i}, v_{\Delta j}\}:_{\substack{i=0,1,2,\ldots,N \\ j=0,N}}$. The cardinal splines have the property that, on the mesh Δ, any cubic function $s_\Delta(x)$ has the representation

$$s_\Delta(x) = \sum_{i=0}^{N} s_\Delta(x^i) w_{\Delta i}(x) + \sum_{j=0,N} \frac{ds_\Delta(x_j)}{dx} v_{\Delta j}(x), \qquad 0 \leq x \leq 1.$$

The cardinal splines $w_{\Delta i}(x)$ $(i = 0, 1, 2, \cdots, N)$ and $v_{\Delta j}(x)$ $(j = 0, N)$ are cubic splines on Δ with

$$w_{\Delta i}(x_k) = \begin{cases} 1 & \text{if } i = k \quad (k = 0, 1, 2, \cdots, N) \\ 0 & \text{otherwise}. \end{cases} \quad i = 0, 1, 2, \cdots, N,$$

$$\frac{\partial w_{\Delta i}(x_k)}{\partial x} = 0, \qquad i = 0, 1, 2, \ldots, N, \quad k = 0, N,$$

$$v_{\Delta i}(x_k) = 0, \qquad j = 0, 1, \ldots, N, \quad i = 0, N,$$

$$\frac{\partial v_{\Delta i}(x_k)}{\partial x} = \begin{cases} 1 & \text{if } k = 0 \text{ or } N \\ 0 & \text{otherwise} \end{cases} \qquad i = 0, N.$$

With these definitions we obtain:

$$w_{\Delta j}(x) = \begin{cases} -\dfrac{2}{h_j^{\,3}}(x-x_j)^3 - \dfrac{3}{h_j^{\,2}}(x-x_j)^2+1, & (x_{j-1} \le x \le x_j), \\[2ex] \dfrac{2}{h_{j+1}^3}(x-x_j)^3 - \dfrac{3}{h_{j+1}^2}(x-x_j)^2+1, & (x_j \le x \le x_{j+1}), \end{cases} \quad j=0,1,2,\ldots,N$$

$$v_{\Delta j}(x) = \begin{cases} \dfrac{1}{h_j^{\,2}}(x-x_j)^3 + \dfrac{2}{h_j}(x-x_j)^2+(x-x_j), & (x_{j-1} \le x \le x_j), \\[2ex] \dfrac{1}{h_{j+1}^2}(x-x_j)^3 - \dfrac{2}{h_{j+1}}(x-x_j)^2+(x-x_j), & (x_j \le x \le x_{j+1}), \end{cases} \quad j=0,N$$

$$h_j = x_j - x_{j-1}, \qquad h_{j+1} = x_{j+1} - x_j\,.$$

A pictorial representation of these cardinal splines is shown in Figure A2.1.

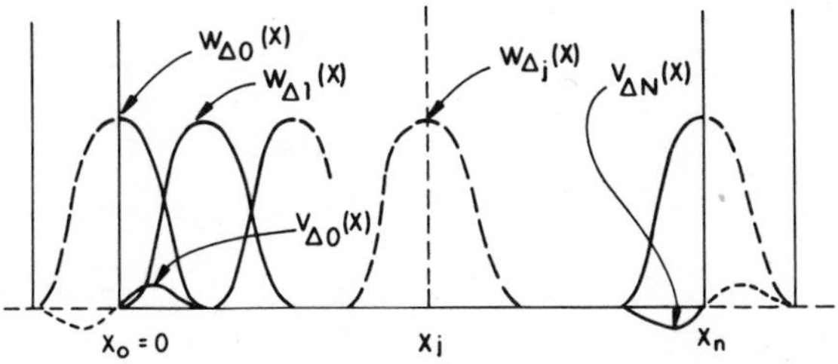

Figure *A*2.1.

We assume a solution to $y(x,t)$ valid in $[0,1]\times[0,T]$:

$$y_N(x,t) = \sum_{j=0}^{N} y(x_j,t)w_{\Delta j}(x) + \sum_{j=0,N} \frac{\partial y(x_j,t)}{\partial x} v_{\Delta j}(x)\,. \tag{i}$$

Note that there are $(N+1)$ unknowns in equation (i) which are to be determined from the requirement that $y_N(x,t)$ satisfy the equations of system I. The unknowns are, of course,

$$y(x_j,t)\,, \qquad j=1,2,\ldots,N-1\,,$$

$$\frac{\partial y(x_j,t)}{\partial x}, \qquad j=0,N\,.$$

Remark. $y(x_0, t) = u_1(0, t)$ and $y(x_N, t) = u_1(1, t)$. It is assumed that analytic expressions are known for $u_1(0, t)$ and $u_1(1, t)$ in what follows. It is this assumption which gives rise to an important objection to the method, of which more will be said later.

We modify equation (2.4) in a manner analogous to that of Section A.3, Chapter 2.

Define

$$\Psi_i(x) = \begin{cases} w_{Ai}(x), & i = 1, 2, \cdots, N-1, \\ v_{Ai}(x), & i = -1, \quad N+1. \end{cases}$$

$$\phi_i(x, t) = g(t)\Psi_i(x), \qquad i \in \{-1, 1, 2, \cdots, N-1, N+1\},$$

$$g(t) \in C^1(0, T), \qquad g(T) = 0.$$

$$y_i(t) = \begin{cases} y(x_i, t), & i = 1, 2, \ldots, N-1, \\ \dfrac{\partial y}{\partial x}(x_0, t), & i = -1, \\ \dfrac{\partial y}{\partial x}(x_N, t), & i = N+1. \end{cases}$$

Then, proceeding as in Section A.3 of Chapter 2, an integral representation of system I is obtained:

$$\int_\Omega \int_0^T \left\{ \left[-\frac{\partial \phi_i}{\partial t} + A[\phi_i(x, t)] \right] y(x, t) - f(x, t)\phi_i(x, t) \right\} dx\, dt$$

$$- \int_\Omega y(x, 0)\phi_i(x, 0)\, dx + \int_\Sigma \left[y(\Sigma)\frac{\partial \phi_i(\Sigma)}{\partial v_{A^*}} - \phi_i(\Sigma)\frac{\partial y(\Sigma)}{\partial v_A} \right] d\Sigma = 0. \qquad \text{(ii)}$$

By definition of $\phi_i(x, t)$, we have

$$\left[y(\Sigma)\frac{\partial \phi_i(\Sigma)}{\partial v_{A^*}} - \phi_i(\Sigma)\frac{\partial y}{\partial v_A} \right] d\Sigma = \begin{cases} \displaystyle\int_0^T [-u_1(0, t)]g(t)\, dt, & i = -1, \\ 0 & i = 1, 2, \ldots, N-1, \\ \displaystyle\int_0^T [u_1(1, t)]g(t)\, dt & i = N+1. \end{cases}$$

Put $y_N(x,t)$ given by (i) into (ii) and adopt the following definitions:

$$f_N(x,t) = \sum_{j=-1}^{N+1} f_j(t)\,\Psi_j(x), \qquad f_j(t) = \frac{\partial f(x_j,t)}{\partial x}, \quad j=-1, N+1,$$

$$u_2(x) = \sum_{j=-1}^{N+1} u_{2j}\,\Psi_j(x), \qquad u_{2j} = \frac{\partial u_2(x_j)}{\partial x}, \quad j=-1, N+1.$$

Remark. $f_j(t)$, u_{2j} ($j=0,1,\cdots,N$) are usually taken to be

$$f_j(t) = f(x_j,t)\,,$$
$$u_{2j} = u_2(x_j,t)\,.$$

Varga [26] has found that an $L^2(\Omega)$ approximation is more accurate:

$$f_j(t) = \frac{1}{\|\Psi_i\|^2_{L^2(\Omega)}} \int_\Omega f(x,t)\,\Psi_j(x)\,dx\,,$$

$$u_{2j} = \frac{1}{\|\Psi_i\|^2_{L^2(\Omega)}} \int_\Omega u_2(x)\Psi_j(x)\,dx\,.$$

Proceeding, we find that (ii) becomes:

$$\sum_{j=-1}^{1} \left\{ \left[\int_0^{h_1} \Psi_{-1}(x)\,\Psi_j(x)\,dx\right]\frac{dy_j}{dt} + \left[\int_0^{h_1} \frac{d\Psi_{-1}(x)}{dx}\,\frac{d\Psi_j(x)}{dx}\,dx\right] y_j(t) \right.$$
$$\left. - \left[\int_0^{h_1} \Psi_{-1}(x)\,\Psi_j(x)\,dx\right] f_j(t) \right\} = 0, \qquad i=-1,$$

$$\sum_{j=i-1}^{i+1} \left\{ \left[\int_{x_{i-1}}^{x_{i+1}} \Psi_i(x)\,\Psi_j(x)\,dx\right]\frac{dy_j}{dt} + \left[\int_{x_{i-1}}^{x_{i+1}} \frac{d\Psi_i(x)}{dx}\,\frac{d\Psi_j(x)}{dx}\,dx\right] y_j(t) \right.$$
$$\left. - \left[\int_{x_{i-1}}^{x_{i+1}} \Psi_i(x)\,\Psi_j(x)\,dx\right] f_j(t) \right\} = 0, \qquad i=1,2,\ldots,N-1,$$

$$\sum_{j=N-1}^{N} \left\{ \left[\int_0^{h_N} \Psi_{N+1}(x)\,\Psi_j(x)\,dx\right]\frac{dy_j}{dt} + \left[\int_0^{h_N} \frac{d\Psi_{N+1}(x)}{dx}\,dx\right] y_j(t) \right.$$
$$\left. - \left(\int_0^{h_N} \Psi_{N+1}(x)\,\Psi_j(x)\,dx\right) f_j(t) \right\} = 0, \qquad i=N+1.$$

This system of equations has the matrix-vector representation:

$$\mathbf{H}\frac{d\mathbf{Y}}{dt} + B\mathbf{Y} - \mathbf{HF} + \mathbf{G} = 0, \qquad \mathbf{H}[\mathbf{Y}(0) - \mathbf{U}] = 0,$$

where

$$\mathbf{H} = \{h_{ij}\}_{i,j=-1,1,2,\dots,N-1,N+1},$$

$$h_{ij} = \begin{cases} \int_0^{h_1} \Psi_{-1}(x)\Psi_j(x)\,dx, & i=-1, \quad j=-1,1, \\ \int_{x_{i-1}}^{x_{i+1}} \Psi_i(x)\Psi_j(x)\,dx, & i=1,2,\dots,N-1, \\ & j=-1,1,2,\dots,N-1,N+1 \\ \int_0^{h_N} \Psi_{N+1}(x)\Psi_j(x)\,dx, & i=N, \quad j=N-1,N+1, \end{cases}$$

$$\mathbf{B} = \{b_{ij}\}_{i,j=-1,1,2,\dots,N-1,N+1},$$

$$b_{ij} = \begin{cases} \int_0^{h_N} \frac{d\Psi_{-1}}{dx}\frac{d\Psi_j}{dx}\,dx, & i=-1, \quad j=-1,1 \\ \int_{x_{i-1}}^{x_{i+1}} \frac{d\Psi_i}{dx}\frac{d\Psi_j}{dx}\,dx, & i=1,2,\dots,N-1, \\ & j=-1,1,2,\dots,N-1,N+1, \\ \int_0^{h_N} \frac{d\Psi_{N+1}}{dx}\frac{d\Psi_j}{dx}\,dx, & i=N, \quad j=N-1,N+1, \end{cases}$$

$$\mathbf{F} = [f_{-1}(t)f_1(t)\cdots f_{N-1}(t)f_{N+1}(t)]^T,$$

$$\mathbf{G} = [g_{-1}(t)g_1(t)\cdots g_{N-1}(t)g_N(t)]^T,$$

$$g_k(t) = \begin{cases} h_{k0}\frac{dy_0}{dt} + b_{k0}y_0 - h_{k0}f_0, & k=-1,1, \\ 0, & k=0,2,\dots,N-2,N, \\ h_{kN}\frac{dy_N}{dt} + b_{kN}y_N - h_{kN}f_N, & k=N-1,N+1, \end{cases}$$

$$\mathbf{Y} = [y_{-1}(t)\,y_1(t)\dots y_{N-1}(t)\,y_{N+1}(t)]^T,$$

$$U = [u_{2,-1}u_{21}u_{22}\dots u_{2N-1}\,u_{2N+1}]^T.$$

H is nonsingular by virtue of the linear independence of the set

$$\{\Psi_i\}_{i=-1,0,1,\ldots,N,N+1}.$$

Thus we can solve for $y(t)$ by solving the $(N+1)$-dimensional ordinary differential equation:

$$\frac{d\mathbf{Y}}{dt}+\mathbf{H}^{-1}\mathbf{B}\mathbf{Y}-\mathbf{F}+\mathbf{H}^{-1}\mathbf{G}=0, \qquad \text{(iii)}$$

$$\mathbf{Y}(0)=\mathbf{U}.$$

Remarks. (i) The vector **G** contains terms such as $(d/dt)u_1(0,t)$, $(d/dt)u_1(1,t)$ which are readily evaluated, provided the analytic form of $u_1(0,t)$ is known. However, in the context of identification or control, $u_1(0,t)$ *is not known analytically* but only as the result of some numerical computation. In that case the numerically unattractive expedient of differentiating the data obtained for $u_1(0,t)$ must be followed. Thus there is a serious objection to using the spline approximation technique in the context of identification or control.

(ii) Having resolved the difficulty of (i), we are faced with the inversion of H and, in addition, a numerically unstable system (iii). The instability is induced by equations for $(\partial y/\partial t)y(x_0,t)$ and $(\partial y/\partial t)y(x_N,t)$, but can be avoided by iterative methods.

(iii) The drawbacks cited in (i) and (ii) could be overcome by the numerical accuracy of the solution, which under appropriate hypothesis [25] is striking. For example, the cubic spline is the interpolate which has minimum curvature between mesh pionts and, moreover, possesses the following error estimates, which are optimal:

$$|f^{(\alpha)}(x)-s_\Delta{}^{(\alpha)}(f;x)|\leq K\|\Delta\|^{2n-\alpha-1}, \qquad \alpha=0,1,\ldots,2n-1,$$

where

$$f^{(\alpha)}(x)=\frac{d^\alpha f(x)}{dx^\alpha},$$

$$\|\Delta\|=\sup_i|x_i-x_{i-1}|,$$

and $s_\Delta(f;x)$ is the cubic spline of interpolation to the function f on the mesh Δ: $a=x_0\leq x_1\cdots\leq x_N=b$.

Thus we conclude that, where accuracy is of paramount importance, the spline approximation is a judicious choice of interpolate. However, if it can be avoided (in the setting of identification), then a saving in computation time may result.

APPENDIX 2.4

CONVEXITY OF A QUADRATIC FUNCTIONAL

The assertion is that

$$J(\lambda\mathbf{u}+(1-\lambda)\mathbf{v}) \leq \lambda J(\mathbf{u})+(1-\lambda)J(\mathbf{v}), \qquad 0 \leq \lambda \leq 1 .$$

Proof.

$$\begin{aligned}(1-\lambda)J(\mathbf{v})+\lambda J(\mathbf{u}) &= (1-\lambda)a(\mathbf{v},\mathbf{v})+\lambda a(\mathbf{u},\mathbf{u})-2[(1-\lambda)L(\mathbf{v})-\lambda L(\mathbf{u})] \\ &= a(\mathbf{v},\mathbf{v})+\lambda a(\mathbf{v},\mathbf{u}-\mathbf{v})+\lambda a(\mathbf{u}-\mathbf{v},\mathbf{v}) \\ &+\lambda a(\mathbf{u}-\mathbf{v},\mathbf{u}-\mathbf{v})-2[L((1-\lambda)\mathbf{v}+\lambda\mathbf{u})] .\end{aligned}$$

Since $a(\mathbf{v},\mathbf{v})$ is coercive,

$$\lambda a(\mathbf{u}-\mathbf{v},\mathbf{u}-\mathbf{v}) \geq \lambda^2 a(\mathbf{u}-\mathbf{v},\mathbf{u}-\mathbf{v}) ,$$

$$\begin{aligned}\therefore (1-\lambda)J(\mathbf{v})+\lambda J(\mathbf{u}) &\geq a(\mathbf{v},\mathbf{v})+\lambda a(\mathbf{v},\mathbf{u}-\mathbf{v})+\lambda a(\mathbf{u}-\mathbf{v},\mathbf{v}) \\ &\quad +\lambda^2 a(\mathbf{u}-\mathbf{v},\mathbf{u}-\mathbf{v})-2L[(1-\lambda)\mathbf{v}+\lambda\mathbf{u}] \\ &= a((1-\lambda)\mathbf{v}+\lambda\mathbf{u}\ (1-\lambda)\mathbf{v}+\lambda\mathbf{u})-2L((1-\lambda)\mathbf{v}+\lambda\mathbf{u}) \\ &= J((1-\lambda)\mathbf{v}+\lambda\mathbf{u}) .\end{aligned}$$

APPENDIX 2.5

PROOF OF UNIQUENESS

We prove the following lemma:

LEMMA 2.1. *If a solution exists for system (I), system (II), or system (III), such that $y(.,\cdot)\in L^2(0,T;H^1)$ then in each case that solution is unique.*

PROOF. Equation (2.4) is linear in $y(x,t)$. Thus it is sufficient to show, that, if

$$\begin{aligned} u_1(.) &= 0\,, \\ u_2(\cdot) &= 0\,, \\ f(\cdot,\cdot) &= 0\,, \text{ then} \\ y(\cdot,\cdot) &= 0\,. \end{aligned}$$

Let $\Psi(\cdot,t)\in L^2(\Omega)$. Multiply equation (2.4) by $\Psi(x,t)$ and integrate over Ω:

$$\int_\Omega \left\{ \frac{\partial y(x,t;\mathbf{0})}{\partial t}\Psi(x,t) + \sum_{i,j=1}^{r} a_{ij}(x,t)\frac{\partial y(x,t,\mathbf{0})}{\partial x_j}\frac{\partial \Psi(x,t)}{\partial x_i} + a_0(x,t)\,y(x,t;\mathbf{0})\,\Psi(x,t)\right\} dx = 0. \tag{i}$$

Let $\Psi(x,t) = y(x,t;0)$. Note that

$$\frac{d}{dt}(y(t),y(t))_{L^2(\Omega)} = \frac{d}{dt}\|y(t)\|^2_{L^2(\Omega)} = 2\left(\frac{\partial y(t)}{\partial t},\, y(t)\right)_{L^2(\Omega)}.$$

Putting this result in (i) and using the hypothesis on the operator $A[\cdot]$, then integration over $(0,T]$ yields:

$$\|y(T)\|^2_{L^2(\Omega)} + \int_0^T \alpha\|y(t)\|^2_{H^1(\Omega)}\,dt \le 0 \Rightarrow y(t) = 0, \qquad t\in(0,T].$$

Remark. In case $y(\cdot,t)\notin H^1(\Omega)$, as can occur for system I under the given hypothesis, then the proof given is not applicable. A proof has been reported by Lions and Magenes [20], and we consider that case separately (Lemma 3.2, Chapter 2).

APPENDIX 2.6

"WELL-SET" PROPERTY

PROOF OF LEMMA 2.8. For **u**, **v**, **h** ∈ V, let $\mathbf{v} = \mathbf{u} + \delta\mathbf{h}$, $\delta \in R^1$. Evidently such a choice of **v** satisfies (2.16). Noting that

$$y(x, t; \mathbf{u}) - y(x, t; \mathbf{v}) = y(x, t; \mathbf{u}) - y(x, t; 0) + y(x, t; 0) - y(x, t; \mathbf{v}) ,$$

then $\|y(x, t; \mathbf{u}) - y(x, t; \mathbf{v})\|_{L^2(Q)}$ is equivalent to

$$\delta^2 \|y(x, t; \mathbf{h}) - y(x, t; \mathbf{0})\|^2_{L^2(Q)} \leq \delta^2 m = \varepsilon$$

since solutions $y(x, t; \mathbf{h})$ exist in $L^2(Q)$ for each system.

APPENDIX 2.7

EXISTENCE AND UNIQUENESS OF EXTREMALS TO $J(\mathbf{v})$

PROOF OF THEOREM 2.3. (i) *Existence.* To show existence of $\mathbf{u} \in K$, we proceed as follows:

(a) Show that a minimizing sequence of elements $\mathbf{v}_n \in K$ is such that

$$\|\mathbf{v}_n\|_V \quad \text{bounded}\,.$$

(a) implies that there exists a subsequence $\mathbf{v}_{n_\mu} \in K$ which converges weakly to $\mathbf{w}$.

(b) Show that $\mathbf{w} \in K$.

(c) Use the lower semicontinuity property of the norm $\|.\|_V$ to deduce that

$$J(\mathbf{w}) = \inf_{\mathbf{v} \in K} J(\mathbf{v})\,.$$

Thus we start by showing (a). First we assume a nontrivial solution, that is, $K \neq \phi$:

$$J(\mathbf{v}) = a(\mathbf{v}, \mathbf{v}) - 2L(\mathbf{v}) \geq \alpha\|\mathbf{v}\|_N{}^2 - 2L(\mathbf{v}) \geq \alpha\|\mathbf{v}\|_N{}^2 - 2|L(\mathbf{v})|\,,$$

$$\therefore J(\mathbf{v}) \geq \alpha\|\mathbf{v}\|_V{}^2 - 2\|L\|\ \|\mathbf{v}\|_V\,,$$

$$\frac{J(\mathbf{v})}{\|\mathbf{v}\|_V} \to \infty \quad \text{as } \|\mathbf{v}\| \to \infty \Rightarrow \|\mathbf{v}_n\|_V \text{ is bounded.}$$

Since $\|\mathbf{v}_n\|_V$ is bounded, there exists a subsequence $\mathbf{v}_{n_\mu}$ such that

$$\mathbf{v}_{n_\mu} \xrightarrow{\text{weakly}} \mathbf{v}_n\,;$$

that is

$$(\mathbf{v}_{n_\mu}, \mathbf{v})_V \to (\mathbf{w}, \mathbf{v})_V \quad \text{for all } \mathbf{v} \in V\,.$$

We show (b) by stating the following lemma:

LEMMA 2.10. *Given K closed and convex, then K is weakly closed.*

Thus, for

$$\mathbf{v}_{n_\mu} \in K, \quad \mathbf{v}_{n_\mu} \xrightarrow{\text{weakly}} \mathbf{w} \Rightarrow \mathbf{w} \in K.$$

The characterization of the lower semicontinuity of the norm $\|.\|_V$ is given in

LEMMA 2.11.

$$\liminf_{\mu \to \infty} \|\mathbf{v}_{n_\mu}\|_V \geq \|\mathbf{w}\|_V \qquad \text{if } \mathbf{v}_{n_\mu} \xrightarrow{\text{weakly}} \mathbf{w}.$$

We now deduce that $\mathbf{w}$ has the minimum property. First we make the following assertion:

Assertion 2.1. Lemma 2.11 implies that

$$\liminf_{\mu \to \infty} J(\mathbf{v}_{n_\mu}) \geq J(\mathbf{w}).$$

PROOF.

$$\begin{aligned} J(\mathbf{v}_{n_\mu}) - J(\mathbf{w}) &= a(\mathbf{v}_{n_\mu}, \mathbf{v}_{n_\mu}) - a(\mathbf{w}, \mathbf{w}) + (\mathbf{f}, \mathbf{v}_{n_\mu} - \mathbf{w}) \\ &\geq \alpha(\|\mathbf{v}_{n_\mu}\|_V{}^2 - \|\mathbf{w}\|_V{}^2) + (\mathbf{f}, \mathbf{v}_{n_\mu} - \mathbf{w}). \end{aligned}$$

Using Lemma 2.11 and the weak convergence of $\mathbf{v}_{n_\mu}$ to $\mathbf{w}$,

$$\liminf_{\mu \to \infty} \{J(\mathbf{v}_{n_\mu}) - J(\mathbf{w})\} \geq 0,$$

$$\liminf_{\mu \to \infty} J(\mathbf{v}_{n_\mu}) \geq J(\mathbf{w}).$$

By hypothesis, $\mathbf{v}_n$ is a minimizing sequence, that is,

$$\liminf_{\mu \to \infty} J(\mathbf{v}_{n_\mu}) \to \inf_{\mathbf{v} \in K} J(\mathbf{v}) = j.$$

By Assertion 2.1,

$$J(\mathbf{w}) \leq j,$$

which is a contradiction unless

$$J(\mathbf{w}) = j = \inf_{\mathbf{v} \in K} J(\mathbf{v}), \qquad \mathbf{w} \in K.$$

The uniqueness follows from the convexity of $J(\mathbf{v})$. We assume the contrary. Namely, take $\mathbf{u}_1$ and $\mathbf{u}_2 \in K$ such that:

$$J(\mathbf{u}_1) = J(\mathbf{u}_2) = j \qquad \mathbf{u}_1 \neq \mathbf{u}_2.$$

By Lemma 2.9, we have

$$J((1-\lambda)\mathbf{u}_1+\lambda\mathbf{u}_2) \leq (1-\lambda)J(\mathbf{u}_1)+\lambda J(\mathbf{u}_2) = j\,.$$

Let

$$(1-\lambda)\mathbf{u}_1+\lambda\mathbf{u}_2 = \mathbf{u}_3 \in K\,.$$

Then

$$J(\mathbf{u}_3) \leq j\,, \quad \text{a contradiction}\,.$$

Thus $\mathbf{u}_1 = \mathbf{u}_2$ and uniqueness follows.

APPENDIX 3.1

THE LUMPED APPROXIMATION TO THE DISTRIBUTED STOCHASTIC MODEL

Distributed System S_D. The distributed system which is approximated by an eigenfunction expansion is the linear parabolic system with Dirichlet boundary conditions:

$$\frac{\partial y(x,t)}{\partial t} + A[y(x,t)] = f(x,t) \qquad \text{in } Q, \tag{i}$$

$$y(\Sigma) = u_1(\Sigma) \qquad \text{on } \Sigma, \tag{ii}$$

$$y(x,0) = u_2(x) \qquad \text{in } \Omega. \tag{iii}$$

Under the hypothesis that:

$$f(\cdot,\cdot) \in L^2(Q) \qquad u_1(\cdot) \in L^2(\Sigma) \qquad u_2(\cdot) \in L^2(\Omega).$$

Then there exists a unique $y(\cdot,\cdot) \in L^2(Q)$. It is known that the eigenfunctions $\{w_i(x)\}_{i=1,2,\ldots}$ generated as solutions to

$$A[w] - \lambda w = 0 \qquad \text{in } \Omega, \tag{iv}$$

$$w = 0 \qquad \text{on } \Gamma \tag{v}$$

are *complete* in $L^2(\Omega)$. Therefore there exists an approximation $y_m(x,t)$ with the property that

$$\lim_{m\to\infty} y_m(x,t) \to y(x,t) \quad \text{almost everywhere in } L^2(Q)$$

with $y_m(x,t)$ given by:

$$y_m(x,t) = \sum_{i=1}^{m} y_i(t) w_i(x). \tag{vi}$$

An equation for the $y_i(t)$, $(i = 1,2,\cdots,m)$ is obtained via the Galerkin technique applied to the system (i), (ii), (iii):

$$\frac{dy_i}{dt} + \lambda_i y_i(t) = f_i(t) - u_{1i}(t) \qquad i = 1,2,\ldots,m, \tag{vii}$$

$$y_i(0) = u_{2i}$$

where

$$f_i(t) = \int_\Omega f(x,t)w_i(x)\,dx, \tag{viii}$$

$$u_{1i}(t) = \int_\Omega u_1(\Sigma)\frac{\partial w_i(\Gamma)}{\partial \nu_{A^*}}\,d\Gamma, \qquad i = 1, 2, \ldots, m, \tag{ix}$$

$$u_{2i}(t) = \int_\Omega u_2(x)w_i(x)\,dx. \tag{x}$$

In addition, there are the measurement processes

$$z_i(t) = \int_\Omega z(x,t)w_i(x)\,dx = y_i(t) + \int_\Omega \varepsilon(x,t)w_i(x)\,dx,$$

$$z_{1i}(t) = \int_\Gamma z_1(\Sigma)\frac{\partial w_i}{\partial \nu_A}\,d\Gamma = u_{1i}(t) + \int_\Gamma \varepsilon_1(\Sigma)\frac{\partial w_i}{\partial \nu_{A^*}}(\Gamma)\,d\Gamma, \quad \text{and}$$

$$z_{2i} = \int_\Omega z_2(x)w_i(x)\,dx = u_{2i} + \int_\Omega \varepsilon_2(x)w_i(x)\,dx \tag{xi}$$

We note that $w_i(.) \in L^2(\Omega)$ and that $\partial w_i/\partial \nu_{A^*}(.) \in L^2(\Gamma)$. Thus $\varepsilon_i(t)$, $\varepsilon_{1\,i}(t)$, and $\varepsilon_{2\,i}$ correspond to the random variables $n(t)$, $n_1(t)$, and $n_2(0)$ defined in Section A.3 of Chapter 3.

Recall that we defined the following properties for these random variables:

$$E[\varepsilon_i(t)] = 0 \qquad \text{for all } t \in (0, T]$$

$$E[\varepsilon_{1\,i}(t)] = 0 \qquad \text{for all } t \in (0, T], \qquad (i = 1, 2, \cdots, m),$$

$$E[\varepsilon_{2\,i}] = \bar{n}_{2\,i}.$$

$$E[\varepsilon_i(t)\varepsilon_j(\tau)] = \Delta_{ij},$$

$$E[\varepsilon_{1i}(t)\varepsilon_{1j}(\tau)] = E\left\{\int_\Gamma u_1(\Sigma)u_1(\Sigma)\frac{\partial w_i}{\partial \nu_{A^*}}\frac{\partial w_j}{\partial \nu_A}\,d\Gamma\right\} = r_{ij}(u_1),$$

$$E[\varepsilon_{2i}\varepsilon_{2j}] = \Delta_{ij},$$

where

$$\Delta_{ij}\begin{cases} 1 & \text{if } i = j, \\ 0 & \text{otherwise}, \end{cases}$$

$$\lim_{m\to\infty} \sum_{i,j=1}^{m} r_{ij}^{-1}\{[u_{1i} - z_{1i}][u_{1j} - z_{1j}]\} = [u_1(\Sigma) - z_1(\Sigma)]^2.$$

Given the system S_D and the measurements $z(t), u_1(t)$, and $u_2(t)$, the problem established was

$$\max_{y(\cdot,\cdot)\in S_D} P[y(\cdot,\cdot)-\Delta(\cdot,\cdot)\le y(\cdot,\cdot)\le y(\cdot,\cdot)+\Delta(\cdot,\cdot)|\, z(\cdot,\cdot)].$$

The approximate problem (in the context of the proceding) is to

$$\max_{y(\cdot,\cdot)\in S_D} P[y_m(\cdot,\cdot)-\Delta_m(\cdot,\cdot)\le y_m(\cdot,\cdot)\le y_m(\cdot,\cdot)+\Delta_m(\cdot,\cdot)|\, z_m(\cdot,\cdot)].$$

Since $w_i(x)$ are deterministic, $i=1,2,\cdots,m$, this is equivalent to

$$\max_{y_i(t)} P[y_i(\cdot)-\Delta_i(\cdot)\le y_i(\cdot)\le y_i(\cdot)+\Delta_i(\cdot)|\, z_i(\cdot)], \qquad i=1,2,\ldots,m,$$

$$= \max_{\substack{y_i\in S_D\\(i=1,2,\ldots,m)}} \iint_{y_i-\Delta_i}^{y_i+\Delta_i}\iint p(y_1,y_2,\ldots,y_m|z_1,z_2,\ldots,z_m)\,dy_1,\ldots,dy_m. \quad \text{(xii)}$$

Note that, by Bayes' rule,

$$p[y_1(\cdot)y_2(\cdot)\ldots y_m(\cdot)|z_1(\cdot)z_2(\cdot)\ldots z_m(\cdot)]$$

$$=\frac{p[z_1(\cdot)z_2(\cdot)\ldots z_m(\cdot)|y_1(\cdot)y_2(\cdot)\ldots y_m(\cdot)]\,p[y_1(\cdot)y_2(\cdot)\ldots y_m(\cdot)]}{p[z_1(\cdot)z_2(\cdot)\ldots z_m(\cdot)]}.$$

Owing to the nonnegativity of the density function,

$$\text{(xii)}\Rightarrow \max_{\substack{y_i(\cdot)\in S_D\\(i=1,2,\ldots,m)}} \{p[z_1z_2\ldots z_m|y_1y_2\ldots y_m]\,p[y_1y_2\ldots y_m]\}. \quad \text{(xiii)}$$

Each of the probabilities in (xiii) is Gaussian. In particular,

$$p[z_1z_2\ldots z_m|y_1y_2\ldots y_m]=k_z\exp -\tfrac{1}{2}\left\{\int_0^T \sum_{i,j=1}^m [y_i(\tau)-z_i(\tau)]^2\,d\tau\right\}, \quad \text{(xiv)}$$

$$p[y_1y_2\ldots y_m]=k_y\exp -\tfrac{1}{2}\left\{\sum_{i,j=1}^m [y_i(0)-z_i(0)]^2\right.$$

$$\left.+\int_0^T \sum_{i,j=1}^m [y_i(\tau)-\bar{y}_i(\tau)]q_{ij}^{-1}[y_i(\tau)-\bar{y}_j(\tau)]\,d\tau\right\}, \quad \text{(xv)}$$

where

$$y_i(\tau)=E\{y_i(\tau)\};\quad q_{ij}=E\{[y_i(\tau)-y_i(\tau)][\bar{y}_j(\tau)-\bar{y}_j(\tau)]\}.$$

Using the relationship

$$y_i(\tau) = e^{-\lambda_i \tau} y_i(0) + \int_0^\tau e^{-\lambda_i(\tau-s)} u_{1i}(\tau)\, d\tau ,$$

we can obtain, via (xv) and (xiv), the following expression for (xiii):

$$\max_{\substack{y_i(\cdot)\in S_D \\ (i=1,2,\ldots,m)}} \quad k_y k_z \exp -\tfrac{1}{2}\left\{\int_0^T \sum_{i,j=1}^m [y_i(\tau)-z_i(\tau)]^2\, d\tau\right.$$

$$+ \int_0^T \sum_{i,j=1}^m [u_{1i}-z_{1i}] r_{ij}^{-1} [u_{ij}-z_{ij}]\, dt$$

$$\left. + \sum_{i,j=1}^m [y_i(0)-z_i(0)][y_j(0)-z_j(0)]\right\}. \qquad \text{(xvi)}$$

Thus we wish to minimize the exponent of (xvi). Note if this exponent is multiplied through by

$$\int_\Omega w_i(x) w_j(x)\, dx = \Delta_{ij} ,$$

then the exponent becomes:

$$\left\{\int_Q [y_m(x,t)-z_m(x,t)]^2\, dx\, dt + \int_\Sigma (u_1-z_1)^2 d\Sigma + \int_\Omega (u_2-z_2)^2\, dx\right\}. \qquad \text{(xvii)}$$

Thus we minimize (xvii) along a trajectory $y_m(x,t) \in S_D$ in order to obtain the maximum likelihood estimate according to Bayes. If we let $m \to \infty$, (xvii) is equivalent to (3.24).

APPENDIX 5.1

EXISTENCE OF AN AFFINE MAP BETWEEN SYSTEM y AND ADJOINT p

Given the system S defined by (5.1), (5.2), and (5.3) the adjoint system S^* defined by (5.4), (5.5), and (5.6), we prove the

Assertion. There exists an affine mapping $\Pi[.]$ defined by

$$\Pi\colon y(\cdot, t; \mathbf{u}) \to p(\cdot, t; \mathbf{u}),$$
$$\Pi\colon L^2(\Omega) \to H_0^{\,1}(\Omega).$$

PROOF. A mapping with the desired properties will be constructed. This construction will be given for the approximate solution to S and S^* denoted by S_m and $S_m{}^*$ respectively. By limiting arguments, the mapping between S and S^* wiil be demonstrated.

Define System S_m.

$$y_m(x, t) = \sum_{i=1}^{m} y_i(t) w_i(x) \quad \text{in } \Omega \times (0, T], \tag{i}$$

where

$$\{w_i(.)\}_{i=1,2,\ldots} \text{ are an orthonormal basis in } L^2(\Omega),$$

and

$y_i(t)$ $(i = 1, 2, \cdots, m)$ is the unique solution of

$$\frac{dy_i(t)}{dt} + \lambda_i y_i(t) = f_i(t) - u_{1i}(t), \quad \text{in } (0, T], \tag{ii}$$

$$y_i(0) = u_{2i}, \qquad t = 0, \tag{iii}$$

$$f_i(t) = \int_\Omega f(x, t) w_i(x)\, dx, \qquad u_{2i} = \int_\Omega u_2(x) w_i(x)\, dx,$$

$$u_{1i}(t) = \int_\Gamma u_1(\Sigma) \frac{\partial w_i}{\partial \nu_{A^*}}\, d\Gamma. \tag{iv}$$

As we have seen, $y_m(x,t)$ defined in this way has the property that

$$\lim_{m\to\infty} y_m(\cdot,\cdot) \to y(\cdot,\cdot) \quad \text{in } L^2(Q).$$

Define Adjoint System S_m^*.

$$p_m(x,t) = \sum_{i=1}^{m} p_i(t) w_i(x) \quad \text{in } \Omega\times(0,T], \tag{v}$$

$\{w_i\}$ being defined as before and $p_i(t)$ being the unique solution of

$$-\frac{dp_i}{dt} + \lambda_i p_i(t) = y_i(t) - z_i(t) \quad i = 1,2,\ldots,m, \tag{vi}$$

$$p_i(T) = 0. \tag{vii}$$

Again, $p_m(x,t)$ has the property that

$$\lim_{m\to\infty} p_m(\cdot,\cdot) \to p(\cdot,\cdot) \quad \text{in } H_0^{\,1}(Q).$$

As we saw, $u_2(x)$ and $u_1(\Sigma)$ are chosen as follows:

$$u_2(x) = z_2(x) - p(x,0;\mathbf{u}), \qquad u_{2i} = z_{2i} - p_i(0),$$

$$u_1(\Sigma) = z_1(\Sigma) + \frac{\partial p(\Sigma;\mathbf{u})}{\partial \nu_{A^*}}, \qquad u_{1i} = z_{1i} + \int_\Gamma \sum_{j=1}^{m}\left[p_j(t)\frac{\partial w_j}{\partial \nu_{A^*}}\right]\frac{\partial w_i}{\partial \nu_{A^*}}\,d\Gamma.$$

Define the following vectors and matrices:

$$\mathbf{y} = [y_1(t)y_2(t)\ldots y_m(t)]^T,$$

$$\mathbf{p} = [p_1(t)p_2(t)\ldots p_m(t)]^T,$$

$$\mathbf{A} = \begin{matrix} \lambda_1 & 0 & \ldots & 0 & 0 \\ 0 & \lambda_2 & \ldots & 0 & 0 \\ \vdots & \vdots & \ddots & \vdots & \vdots \\ 0 & 0 & \ldots & 0 & \lambda_m \end{matrix},$$

$$\mathbf{W}\mathbf{W}^T = \int_{\Gamma_s} \begin{matrix} \frac{\partial w_1(\Gamma_s)}{\partial \nu_{A^*}} & \frac{\partial w_1(\Gamma_s)}{\partial \nu_{A^*}} & \ldots & \frac{\partial w_1(\Gamma_s)}{\partial \nu_{A^*}} & \frac{\partial w_m(\Gamma_s)}{\partial \nu_{A^*}} \\ \vdots & \vdots & & \vdots & \vdots \\ \frac{\partial w_m(\Gamma_s)}{\partial \nu_{A^*}} & \frac{\partial w_1(\Gamma_s)}{\partial \nu_{A^*}} & \ldots & \frac{\partial w_m(\Gamma_s)}{\partial \nu_{A^*}} & \frac{\partial w_m(\Gamma_s)}{\partial \nu_{A^*}} \end{matrix} \, d\Gamma_s,$$

$$\mathbf{f} = [f_1(t)f_2(t)\ldots f_m(t)]^T, \qquad \mathbf{z}_2 = [z_{21}z_{22}\ldots z_{2m})^T$$
$$\mathbf{z} = [z_1(t)z_2(t)\ldots z_m(t)]^T, \qquad \mathbf{z}_1 = [z_{11}(t)z_{12}(t)\ldots z_{1m}(t)]^T.$$

Then the system of equations (i), (ii), (iii), (v), (vi) and (vii) can be written as:

$$\frac{d\mathbf{y}(t)}{dt} + \mathbf{A}\mathbf{y}(t) = \mathbf{f}(t) - \mathbf{W}\mathbf{W}^T\mathbf{p}(t), \tag{viii}$$

$$\mathbf{y}(0) = \mathbf{z}_2 - \mathbf{p}(0), \tag{ix}$$

$$-\frac{d\mathbf{p}(t)}{dt} + \mathbf{A}\mathbf{p}(t) = \mathbf{y}(t) - \mathbf{z}(t), \tag{x}$$

$$\mathbf{p}(T) = \mathbf{0}. \tag{xi}$$

Define $\mathbf{q}(t) = [\mathbf{y}(t)^T \,\vdots\, \mathbf{p}(t)^T]^T$ ($2m \times 1$ vector):

$$\mathbf{B} = \begin{bmatrix} \mathbf{A} & \mathbf{W}\mathbf{W}^T \\ \mathbf{I} & -\mathbf{A} \end{bmatrix},$$

$$\mathbf{F}(t) = [\mathbf{f}(t)^T \,\vdots\, \mathbf{z}(t)^T]^T.$$

Then (viii) and (x) imply that

$$\frac{d\mathbf{q}}{dt} + B\mathbf{q} = \mathbf{F}(t). \tag{xii}$$

If we define the state transition matrix $\mathbf{\Phi}(t)$ for the time-invariant system (xii) by

$$\frac{d\mathbf{\Phi}}{dt} + \mathbf{B}\mathbf{\Phi} = 0 \qquad \mathbf{\Phi}(t) = \begin{bmatrix} \phi_{11}(t) & \phi_{12}(t) \\ \phi_{21}(t) & \phi_{22}(t) \end{bmatrix}.$$
$$\mathbf{\Phi}(0) = \mathbf{I}$$

Then $\mathbf{q}(t)$ has the following solution:

$$\mathbf{q}(t) = \mathbf{\Phi}(t)\mathbf{q}(0) + \int_0^t \mathbf{\Phi}(t-\tau)\mathbf{F}(\tau)\,d\tau. \tag{xiii}$$

Using the definition of $\mathbf{q}$, (xiii) implies that

$$\mathbf{y}(t) = \phi_{11}(t)\,\mathbf{y}(0) + \phi_{12}(t)\,\mathbf{p}(0) + \int_0^t \{\phi_{11}(t-\tau)\,\mathbf{f}(\tau) + \phi_{12}(t-\tau)\mathbf{z}(\tau)\}\,d\tau, \tag{xiv}$$

$$\mathbf{p}(t) = \phi_{21}(t)\,\mathbf{y}(0) + \phi_{22}(t)\,\mathbf{p}(0) + \int_0^t \{\phi_{21}(t-\tau)\,\mathbf{f}(\tau) + \varphi_{22}(t-\tau)\,\mathbf{z}(\tau)\}\,d\tau. \tag{xv}$$

Using (ix) in (xv) where appropriate, we obtain:

$$\mathbf{y}(t) = [\phi_{12}(t) - \phi_{11}(t)]\,\mathbf{p}(0) + \phi_{11}(t)\,\mathbf{z}_2 + \int_0^t \{\phi_{11}(t-\tau)\,\mathbf{f}(\tau) + \phi_{12}(t-\tau)\,\mathbf{z}(\tau)\}\,d\tau\,, \tag{xvi}$$

$$\mathbf{p}(t) = [\phi_{22}(t) - \phi_{21}(t)]\,\mathbf{p}(0) + \phi_{21}(t)\,\mathbf{z}_2 + \int_0^t \{\phi_{21}(t-\tau)\,\mathbf{f}(\tau) + \phi_{22}(t-\tau)\,\mathbf{z}(t)\}\,d\tau\,. \tag{xvii}$$

Assuming formally that $[\phi_{12}(t) - \phi_{11}(t)]$ has an inverse for all $t \in (0, T]$, then using (xvi) an expression for $\mathbf{p}(0)$ in terms of $\mathbf{y}(t)$ can be obtained:

$$\mathbf{p}(0) = [\phi_{12}(t) - \phi_{11}(t)]^{-1}\,\mathbf{y}(t) - [\phi_{12}(t) - \phi_{11}(t)]^{-1}\{\phi_{11}(t)\,\mathbf{z}_2 + \int_0^t [\phi_{11}(t-\tau)\,\mathbf{f}(\tau) + \phi_{12}(t-\tau)\,\mathbf{z}(\tau)]\}\,d\tau\,. \tag{xviii}$$

Define

$$\mathbf{h}(t) = \phi_{21}(t)\,\mathbf{z}_2 + \int_0^t \{\phi_{21}(t-\tau)\,\mathbf{f}(\tau) + \varphi_{22}(t-\tau)\,\mathbf{z}(\tau)\}\,d\tau - [\phi_{22}(t) - \phi_{21}(t)]\,[\varphi_{12}(t) - \phi_{11}(t)]^{-1}\{\phi_{11}(t)\,\mathbf{z}_2 + \int_0^t [\phi_{11}(t-\tau)\,\mathbf{f}(\tau) + \phi_{12}(t-\tau)\,\mathbf{z}(\tau)]\}\,d\tau\,. \tag{xix}$$

Then, if we replace $\mathbf{p}(0)$ in (xvii) by (xviii) and use the definition of $\mathbf{h}(t)$ afforded by (xix), we obtain finally:

$$\mathbf{p}(t) = [\phi_{22}(t) - \phi_{21}(t)][\phi_{12}(t) - \phi_{11}(t)]^{-1}\mathbf{y}(t) + \mathbf{h}(t)\,. \tag{xx}$$

The finite dimensional result has thus been obtained. Now define

$$\mathbf{P}(t) = [\phi_{22}(t) - \phi_{21}(t)][\phi_{12}(t) - \phi_{11}(t)]^{-1}\,,$$
$$\mathbf{w} \quad = [w_1(x)w_2(x)\ldots w_m(x)]^T\,.$$

Then

$$p_m(x, t) = \mathbf{w}^T(x)\mathbf{p}(t) = \mathbf{w}^T(x)\mathbf{P}(t)\mathbf{y}(t) + \mathbf{w}^T(x)\mathbf{h}(t)\,. \tag{xxi}$$

In summation notation, (xxi) is:

$$p_m(x,t) = \sum_{i=1}^{m} p_i(t)w_i(x) = \sum_{i=1}^{m} \left\{ \sum_{j=1}^{m} P_{ij}(t)\, y_j(t) w_i(x) + h_i(t) w_i(x) \right\}. \tag{xxii}$$

We now anticipate the desired result and write

$$p(x,t) = \int_\Omega P(x,\xi,t)\, y(\xi,t)\, d\xi + h(x,t), \tag{xxiii}$$

$$p_m(x,t) = \int_\Omega [P(x,\xi,t)\, y(\xi,t)]_m\, d\xi + h_m(x,t). \tag{xxiv}$$

By hypothesis (b) of Theorem 5.1,

$$P(\cdot,\cdot,t) \in H^2(\Omega\times\Omega) \Rightarrow P(\cdot,\cdot,t) \in L^2(\Omega\times\Omega).$$

Thus (xxiv) becomes

$$p_m(x,t) = \int_\Omega \left\{ \sum_{i=1}^{m} \sum_{i=1}^{m} P_{ij}(t) w_i(x) w_j(\xi) \sum_{k=1}^{m} y_k(t) w_k(\xi) \right\} d\xi + h_m(x,t),$$

$$\therefore p_m(x,t) = \sum_{i=1}^{m} \left\{ \sum_{i=1}^{m} P_{ij}(t)\, y_j(t) w_i(x) \right\} + h_m(x,t). \tag{xxv}$$

Equation (xxv) is identical to (xxii) with

$$h_m(x,t) = \sum_{i=1}^{m} h_i(t) w_i(x)$$

going to the limit with m in (xxii), then we obtain

$$p(x,t) = \int_\Omega P(x,\xi,t)\, y(\xi,t)\, d\xi + h(x,t).$$

APPENDIX 5.2

DERIVATION OF EQUATIONS FOR $P(x, \xi, t)$ AND $\hat{y}(\xi, t)$

We start with equation (5.29):

$$\int_{\Omega} \left\{ \frac{\partial P(x, \xi, t)}{\partial t} [y(\xi, t; \mathbf{u}) - \hat{y}(\xi, t)] + P(x, \xi, t) \left[\frac{\partial y(\xi, t; \mathbf{u})}{\partial t} - \frac{\partial \hat{y}(\xi, t)}{\partial t} \right] \right.$$

$$\left. - A_x[P(x, \xi, t)] [y(\xi, t; \mathbf{u}) - \hat{y}(\xi, t)] \right\} d\xi = y(x, t; \mathbf{u}) - z(x, t). \qquad \text{(i)}$$

Using (5.25), (5.26), (5.27), and (5.28) along with (5.1) and (5.7), we show that (5.29) (given as (i) here) implies and is implied by (5.30), (5.31), and (5.32). To show the implications, we generate the identities:

(a) $$P(x, \xi, t) \left[\frac{\partial y(\xi, t; \mathbf{u})}{\partial t} - \frac{\partial \hat{y}(\xi, t)}{\partial t} \right] = -P(x, \xi, t) \{ A_\xi[y(\xi, t; \mathbf{u}) - \hat{y}(\xi, t)]$$

$$+ A[\hat{y}(\xi, t)] + \frac{\partial \hat{y}(\xi, t)}{\partial t} - f(\xi, t) \Big\}, \qquad \text{(ii)}$$

where in the LHS of (ii) use was made of the definition of $[\partial y/\partial t(\xi, t; \mathbf{u})]$ afforded by (5.1). Integrating (a) over Ω and using Green's theorem where appropriate, obtain:

(b) $$\int_{\Omega} P(x, \xi, t) \left[\frac{\partial y(\xi, t; \mathbf{u})}{\partial t} - \frac{\partial \hat{y}(\xi, t)}{\partial t} \right] d\xi$$

$$= \int_{\Omega} - \left\{ A_\xi[P(x, \xi, t)] [y(\xi, t; \mathbf{u}) - \hat{y}(\xi, t)] \right.$$

$$\left. + P(x, \xi, t) \left[\frac{\partial \hat{y}(\xi, t)}{\partial t} + A_\xi[\hat{y}(\xi, t)] - f(\xi, t) \right] \right\} d\xi$$

$$+ \int_{\Gamma_\xi} \left\{ P(x, \Sigma_\xi) \left[\frac{\partial y(\Sigma; \mathbf{u})}{\partial \nu_{A_\xi}} - \frac{\partial y(\Sigma)}{\partial \nu_{A_\xi}} \right] - \frac{\partial P}{\partial \nu_{A_\xi^*}} [u_1(\Sigma) - y(\Sigma)] \right\} d\Gamma_\xi .$$

We note that the second to last term in (b) is zero because $P(x, \Sigma_\xi) = 0$ (equation 5.25). The last term of (b) can be phrased in terms of the variables of interest, if we first note that (5.7) implies:

$$\text{(c)} \qquad u_1(\Sigma_x) = z_1(\Sigma_x) - \int_\Omega \frac{\partial P(\Sigma_x, \xi)}{\partial v_{A_x^*}} [y(\xi, t; \mathbf{u}) - \hat{y}(\xi, t)]\, d\xi\,.$$

Putting(c) into the last term of (b) and relabeling the dummy arguments of integration, then:

$$\text{(d)} \qquad -\int_{\Gamma_\xi} \frac{\partial P(x, \Sigma_\xi)}{\partial v_{A_\xi^*}} [u_1(\Sigma_\xi) - z_1(\Sigma_\xi)]\, d\Gamma_\xi$$

$$= \int_\Omega \int_{\Gamma_s} \left\{ \frac{\partial P(x, \Sigma_s)}{\partial v_{A_s^*}} \frac{\partial P(\Sigma_s, \xi)}{\partial v_{A_s^*}} [y(\xi, t; \mathbf{u}) - \hat{y}(\xi, t)] \right\} d\xi\, d\Gamma_s$$

$$- \int_{\Gamma_s} \frac{\partial P(x, \Sigma_s)}{\partial v_{A_s^*}} [z_1(\Sigma_s) - \hat{y}(\Sigma_s)\, d\Gamma_s\,.$$

Putting (d) into (b) gives, on collection of terms, the following:

$$\text{(e)} \qquad \int_\Omega P(x, \xi, t) \left[\frac{\partial y(\xi, t; \mathbf{u})}{\partial t} - \frac{\partial y(\xi, t)}{\partial t} \right] d\xi = \int_\Omega \left\{ \left[-A_\xi[P(x, \xi, t)] \right. \right.$$

$$\left. + \int_{\Gamma_s} \frac{\partial P(x, \Sigma_s)}{\partial v_{A_s^*}} \frac{\partial P(\Sigma_s, \xi)}{\partial v_{A_s^*}} d\Gamma_s \right] [y(\xi, t; \mathbf{u}) - \quad (\xi, t)]$$

$$\left. + P(x, \xi, t) \left[\frac{\partial \hat{y}(\xi, t)}{\partial t} + A[\hat{y}(\xi, t)] - f(\xi, t) \right] \right\} d\xi$$

$$- \int_{\Gamma_s} \frac{\partial P(x, \Sigma_s)}{\partial v_{A_s^*}} [z_1(\Sigma_s) - \hat{y}(\Sigma_s)\, d\Gamma_s\,.$$

Finally, replace the second term of (i) by (e) and write the RHS of (i) as

$$\text{(f)} \quad y(x, t; \mathbf{u}) - z(x, t) = \int_\Omega [y(\xi, t; \mathbf{u}) - \hat{y}(\xi, t)] \delta(\xi - x)\, d\xi + [\hat{y}(x, t) - z(x, t)]\,.$$

Then,

$$\int_{\Omega}\left\{\frac{\partial P(x,\xi,t)}{\partial t} - A_{\xi}[P(x,\xi,t)] - A_{x}[P(x,\xi,t)] - \delta(\xi - x)\right.$$

$$\left. + \int_{\Gamma_s} \frac{\partial P(x,\Sigma_s)}{\partial \nu_{A_s^*}} \frac{P(\Sigma_s,\xi)}{\partial \nu_{A_s^*}}\, d\Gamma_s\right\}[y(\xi,t;\mathbf{u}) - \hat{y}(\xi,t)]\, d\xi$$

$$+ \int_{\Omega} P(x,\xi,t)\left[\frac{\partial \hat{y}(\xi,t)}{\partial t} + A_{\xi}[\hat{y}(\xi,t)] - f(\xi,t)\right] d\xi$$

$$- \int_{\Gamma_s} \frac{\partial P(x,\Sigma_s)}{\partial \nu_{A_s^*}} [\hat{y}(\Sigma_s) - z_1(\Sigma_s)]\, d\Gamma_s = [\hat{y}(x,t) - z(x,t)]. \tag{iii}$$

We define

$$\hat{y}(\Sigma_s) = z_1(s)\,, \tag{iv}$$

$$\int_{\Omega} P(x,\xi,t)\left[\frac{\partial \hat{y}(\xi,t)}{\partial t} + A_{\xi}[\hat{y}(\xi,t)] - f(\xi,t)\right] d\xi = \hat{y}(x,t) - z(x,t). \tag{v}$$

Then (ii) implies that

$$\frac{\partial P(x,\xi,t)}{\partial t} - A_{\xi}[P(x,\xi,t)] - A_{x}[P(x,\xi,t)] - \delta(\xi - x)$$

$$+ \int_{\Gamma_s} \frac{\partial P(x,\Sigma_s)}{\partial \nu_{A_s^*}} \frac{\partial P(\Sigma_s,\xi)}{\partial \nu_{A_s^*}}\, d\Gamma_s = 0. \tag{vi}$$

Equations (iv), and (vi), (c), are the same as (5.32), (5.31) and (5.30); thus it has been shown that (5.29) implies (5.32), (5.31), and (5.30).

The reverse implication follows immediately by retracing the steps (f) through (a).

APPENDIX 5.3

THE VARIATIONAL DERIVATIVE

In Section A.1.3 of Chapter 5, we considered extremals of the quadratic functional $a(\mathbf{v},\mathbf{v})-2L(\mathbf{v})+c$, and determined that the extremal $\mathbf{u}$ to this functional was characterized by $a(\mathbf{u},\mathbf{v})-L(\mathbf{v})=0$ for all $\mathbf{v}\in V$.

Here we shall give an interpretation to that characterization. The notation and terminology used is that of [28]. We have

$$J(\mathbf{v}) = a(\mathbf{v},\mathbf{v})-2L(\mathbf{v})+c, \qquad \mathbf{v}\in V, \tag{i}$$

$$J(\mathbf{v}):\quad V\to R_1, \qquad V = L^2(\Sigma)\times L^2(\Omega).$$

Define the *increment* $\Delta J(\mathbf{h})$ of the functional $J(\mathbf{v})$ by

$$\Delta J(\mathbf{h}) = J(\mathbf{v}+\mathbf{h})-J(\mathbf{v}), \qquad \mathbf{h}\in V;$$

suppose that

$$\Delta J(\mathbf{h}) = \phi(\mathbf{h})+\varepsilon\|\mathbf{h}\|_V^2,$$

where $\phi(\mathbf{h})$ is a linear functional on V and $\varepsilon\to 0$ as $\|\mathbf{h}\|^2\to 0$. Then $J(\mathbf{v})$ is differentiable and the principal linear part of the increment $\Delta J(\mathbf{h})$, that is, the linear functional $\phi(\mathbf{h})$, is called the *variation* or *differential* of $J(\mathbf{h})$ and is denoted by $\delta J(\mathbf{h})$.

We now construct the differential evaluated at $\mathbf{u}$ of the fuctional $J(\mathbf{v})$ given by (i). Let

$$\mathbf{h}=\varepsilon\mathbf{v}, \quad \mathbf{w}=\frac{\varepsilon}{2}\mathbf{v}; \qquad \varepsilon\in R^1;\ \mathbf{v}\in V;\ \mathbf{h},\ \mathbf{w}\in V.$$

Note that

$$\lim_{\varepsilon\to 0} a(\mathbf{h},\mathbf{h})\to 0, \qquad (a(\mathbf{v},\mathbf{v}) \text{ is continuous}).$$

Using (i), we have that

$$\begin{aligned}\Delta J(\mathbf{h}) &= a(\mathbf{u}+\mathbf{h},\mathbf{u}+\mathbf{h})-2L(\mathbf{u}+h)-a(\mathbf{u},\mathbf{u})+2L(\mathbf{u})\\ &= 2a(\mathbf{u},\mathbf{h})-2L(\mathbf{h})+a(\mathbf{h},\mathbf{h}).\end{aligned} \tag{ii}$$

Thus the differential $\delta J(\mathbf{u})$, which is the principal linear part of (ii), is given by

$$\Delta J(\mathbf{w}) = a(\mathbf{u}, \mathbf{w}) - L(\mathbf{w}) . \tag{iii}$$

Equation (iii) is evidently the same as equation (5.10) of Chapter 5, Section A.1 Consequently, the equation characterizing the optimal $\mathbf{u}$ is a statement of the fact the differential of the functional $J(\mathbf{u})$ vanishes.

The variational derivative $\mathbf{G}(\mathbf{u})$ or gradient of the functional $J(\mathbf{u})$ is defined by the relation

$$\Delta J(\mathbf{w}) = (\mathbf{G}(\mathbf{u}), \mathbf{w})_V . \tag{iv}$$

This definition is unique since, by the Riez representation theorem for continuous linear functionals in a Hilbert space, (iii) and (iv) are equivalent.

APPENDIX 5.4

DERIVATION OF SOME IDENTITIES

In the notation of Chapter 5, Section A.3, we can construct the following identities given in Chapter 5, Section A.4.

(a)
$$f_i(t) = \int_\Omega f(x,t)\, w_i(x)\, dx = \int_0^1 212.0\sqrt{2}\sin(i\Pi)x\, dx$$

$$= \frac{212.0\sqrt{2}}{(i\Pi)}[1.0-(-1.0)^i].$$

(b)
$$z_{1i}(t) = \int_\Gamma z_1(\Sigma)\frac{\partial w_i(\Gamma)}{\partial v_{A^*}}\, d\Gamma = \int_\Gamma z_1(\Gamma, t)\frac{\partial w_i(\Gamma)}{\partial v_{A^*}}\, d\Gamma$$

$$= -z_1(0,t)\frac{\partial}{\partial x}[\sqrt{2}\sin(i\Pi)x]_{x=0} + z_1(1,t)\frac{\partial}{\partial x}[\sqrt{2}\sin(i\Pi)x]_{x=1}.$$

$$\therefore z_{1i} = \sqrt{2}(i\Pi)[-(70+10\sin 2\Pi t + k_1 N_1(t)) + (-1)^i(54.5+k_2 N_2(t))].$$

(c)
$$z_{2i} = \int_\Omega z_2(x)\, w_i(x)\, dx = \int_0^1 [70e^{-0.25x} + k_3]\sqrt{2}\sin(i\Pi)x\, dx$$

$$= 70\sqrt{2}\left\{e^{-0.25x}\frac{[-0.25\sin i\Pi x - i\Pi\cos i\Pi x]}{(0.25)^2+(i\Pi)^2}\right\}_0^1 + \frac{k_3\sqrt{2}}{(i\Pi)}[1-(-1)^i]$$

$$= 70\sqrt{2}\left\{\frac{-(i\Pi)(-1)^i e^{-0.25} + (i\Pi)}{(0.25)^2+(i\Pi)^2}\right\} + \frac{k_3\sqrt{2}}{(i\Pi)}[1.0-(-1.0)^i].$$

$$\therefore z_{2i} = \frac{70\sqrt{2}}{(i\Pi)}[1.0-(-1.0)^i e^{-0.25}] + \frac{k_3\sqrt{2}}{(i\Pi)}[1.0-(-1.0)^i].$$

(d) $z_i(t) = \int_\Omega z(x,t)\, w_i(x)\, dx$

$$= \int_0^1 [y(x,t;\mathbf{u}^*) + k_0 N_0(t)] \sqrt{2} \sin(i\Pi)x\, dx$$

$$= y_i(t,\mathbf{u}^*) + \frac{k_0 N_0(t)\sqrt{2}}{(i\Pi)} [1.0 - (-1.0)^i].$$

$y_i(t;\mathbf{u}^*)$ satisfies the following differential equation:

$$\frac{dy_i(t;\mathbf{u}^*)}{dt} - \lambda_i y_i(t;\mathbf{u}^*) = f_i(t) - u_{1i}^*(t),$$

$$y_i(0) = u_{2i}^*.$$

The solution of this equation is given by:

$$y_i(t;\mathbf{u}^*) = e^{-\lambda_i t} u_{2i}^* + \int_0^t e^{-\lambda_i(t-\tau)} [f_i(\tau) - u_{1i}^*(\tau)]\, d\tau,$$

performing the integration,

$$y_i(t;\mathbf{u}^*) = \frac{70\sqrt{2}}{i\Pi} [1.0 - (-1.0)^i e^{-0.25}] e^{-(i\Pi)^2 t} + \left\{ \frac{212.0\sqrt{2}}{(i\Pi)^3} [1.0 - (-1.0)^i] \right.$$

$$\left. + \frac{70\sqrt{2}}{(i\Pi)} - \frac{54.5(-1)^i\sqrt{2}}{(i\Pi)} \right\} \{1.0 - e^{-(i\Pi)^2 t}\}$$

$$+ \frac{10\sqrt{2}(i\Pi)}{(i\Pi)^4 + 4\Pi^2} [(i\Pi)^2 \sin 2\Pi t - 2\Pi \cos 2\Pi t + 2\Pi e^{-(i\Pi)^2 t}].$$

APPENDIX 5.5

CONVERGENCE OF CONJUGATE GRADIENT ALGORITHM

THEOREM 5.2. *Given the hypothesis of Section* A.1.3, *then, if* $\mathbf{G}(\mathbf{u}^i) \neq 0$, $J(\mathbf{u}^{i+1}) < J(\mathbf{u}^i)$.

PROOF. Assume there is no $\alpha > 0$ such that

$$J(\mathbf{u}^{i+1}) = J(\mathbf{u}^i + \alpha \mathbf{S}^i) < J(\mathbf{u}^i) . \tag{i}$$

Hence, for all $\alpha > 0$,

$$\frac{J(\mathbf{u}^i + \alpha \mathbf{S}^i) - J(\mathbf{u}^i)}{\alpha} \geq 0 . \tag{ii}$$

In the limit as α approaches zero, (ii) gives

$$(\mathbf{G}(\mathbf{u}^i), \mathbf{S}^i)_V \geq 0 .$$

$$\therefore -(\mathbf{G}(\mathbf{u}^i), \mathbf{G}(\mathbf{u}^i))_V + \beta({}^i\mathbf{G}(\mathbf{u}^i), \mathbf{S}^{i-1}) \geq 0 . \tag{iii}$$

But (i) implies

$$\frac{d}{d\alpha} J(\mathbf{u}^i + \alpha \mathbf{S}^i) = 0 = (\mathbf{G}(\mathbf{u}^i), \mathbf{S}^{i-1})_V = 0 .$$

Thus (iii) gives a contradiction, namely,

$$-(\mathbf{G}(\mathbf{u}^i), \mathbf{G}(\mathbf{u}^i))_V \geq 0, \qquad \mathbf{G}(\mathbf{u}^i) \neq 0 .$$

Hence there is an α which gives $J(\mathbf{u}^{i+1}) < J(\mathbf{u}^i)$. Since α is chosen according to (i), the theorem is proved.

REFERENCES

1. J. L. Lions, *Contrôle Optimal de Systèmes Gouvernés par des Equations aux Dérivées Partielles*, Dunod, Paris, 1968.
2. A. V. Balakrishnan and J. L. Lions, State estimation for infinite dimensional systems, *J. Comp. System Sci.*, *1* (1967), 391–403.
3. P. K. C. Wang, Control of Distributed Parameter Systems, in *Advances in Control Systems: Theory and Applications*, Academic Press, New York, 1964, pp. 75–172.
4. A. G. Butkovskii and A. Ya. Lerner, The optimal control of systems with distributed parameters, *Automation and Remote Control, 21* (1961), 13–21.
5. A. G. Butkovskii, The maximum principle for systems with distributed parameters, *Automation and Remote Control, 22* (1962), 1156–1169.
6. J. L. Lions, Control problems in systems described by partial differential equations in *Mathematical Theory of Control*, Academic Press, New York, 1967.
7. H. Erzberger, M. Kim, Optimum boundary control of distributed parameter systems, *Information and Control*, (1966), pp. 265–278.
8. T. Kitamore, *Transformation of Distributed Parameter Systems into Lumped Parameter Systems for the Studies of Optimum Control*, I. F. A. C., London, 1966, pp. 24A.3–24A.9.
9. R. S. Bucy, *Nonlinear Filtering Theory*, IEEE Series *G*, Correspondence Item, 1965, p. 198.
10. H.J. Kushner, On the differential equations satisfied by conditional probability density of Markov processes, with applications, *J. S.I.A.M. Control, Ser. A*, 2 (1964), No. 1.
11. W. M. Wonham, Some applications of stochastic differential equations to nonlinear filtering, *J. S.I.A.M. Control, Ser. A*, *2* (1965), No. 3.
12. P. Falb, Infinite dimensional filtering—the Kalman-Bucy filter in Hilbert space, *International Jo. of Control*, 11 (1967), 102–103,
13. A. Bensoussan, *Sur Identification et le Filtrage de Systemes Gouvernés par des Équations aux Dérivées Partielles*, IRIA, Paris, France.
14. J. L. Lions G. Stampacchia Variational inequalities, *Comm. Pure Appl. Math., to* appear.
15. J. W. Dettman, *Mathematical Methods in Physics and Engineering*, McGraw-Hill, New York, 1962.
16. J. L. Lions Functional analysis and optimization, (A two week short course, July 31–August 11, 1967, at U.C.L.A.).
17. L. S. Pontryagin *et al.*, *The Mathematical Theory of Optimal Processes*, New York and Wiley, 1962.
18. A. N. Tikhonov A. A. Samarskii *Equations of Mathematical Physics*, Macmillan, New York, 1963.
19. J. D. Pearson, On the duality between estimation and control, *J. S.I.A.M. Control, Ser. A* (1966).
20. J. L. Lions and E. Magenes, Remarques sur les problèmes aux limites pour opérateurs paraboliques, *C. R. Acad. Sci., Paris, 25* (1960), 2118–2120.
21. R. Fletcher C. M. Reeves Function minimization by conjugate gradients, *Brit. Computer J.*, (July 1964), 149–154.
22. L. S. Lasdon, S. K. Mitter, A. D. Warren, The conjugate gradient method for optimal control problems, *IEEE Ser. G, AC-12* (1968), No. 2, 132–135, 1968.

23. H. A. Antosiewicz, W. C. Rheinboldt, *Numerical Analysis and Functional Analysis*, in, *Survey of Numerical Analysis*, J. Todd Ed., McGraw-Hill, New York, 1962.
24. R. Lattès and J.-L. Lions, *Méthode de Quasi-Réversibilité et Applications*, Dunod, Paris, 1967. (English translation by Richard Bellman, published by American Elsevier, New York, 1969).
25. J. H. Ahlberg, E. N. Nilson and J. L. Walsh, *The Theory of Splines and Their Applications*, Academic Press, New York, 1967.
26. M. H. Shultz and R. S. Varga, *L-Splines*, *Numer. Math. 10* (1967), 345–369.
27. R. E. Kalman, "*New Methods and Results in Linear Prediction and Filtering Theory*", *R.I.A.S.*, Technical Report No. 61-1, 1960.
28. I. M. Gelfand, S. V. Fomin, *Calculus of variations*, Prentice Hall, N.J., 1963.

AUTHOR INDEX

Numbers in parentheses indicate the numbers of the references when these are cited in the text without the names of the authors.

Numbers set in *italics* designate the page numbers on which the complete literature citation is given.

SUBJECT INDEX